Fantastic Journeys

Fantastic Journeys

Five Great Quests of Modern Science

BY MARK HUNTER

WALKER AND COMPANY / New York

Copyright © 1980 by MARK HUNTER

All rights reserved. No part of this book may be reproduced or transmitted in any form or by any means, electric or mechanical, including photocopying, recording, or by any information storage and retrieval system, without permission in writing from the Publisher.

First published in the United States of America in 1980 by the Walker Publishing Company, Inc.

Published simultaneously in Canada by Beaverbooks, Limited, Pickering, Ontario.

ISBN: 0-8027-0638-X

Library of Congress Catalog Card Number: 79-52623

Printed in the United States of America

10 9 8 7 6 5 4 3 2

Book designed by LENA FONG HOR

*To my wife,
without whose love and encouragement
this book would never have been written*

Contents

Foreword xi

Part I—THE EARTH xiii

1. **The Journey to the Center of the Earth** 1
    The Puzzle of the Double Tremor *3*
    The Clue of the Stone Icicles *9*
    Project Mohole *10*
    The Clue of the Undersea Map *12*
    The Clue of the Undersea Volcanoes *14*
    The Clue of the Hot Ice *15*
    The Clue of the Flip-flopping Compass Needles *16*
    The Mystery of the Shadow Zones *22*
2. **The Journey to the Lost Continent** 27
    The Mystery of the Missing Land Bridges *29*
    The Clue of the Aging Volcanic Islands *31*
    The Puzzle of the Deep-Sea Zebra Stripes *35*
    The Clue of the Mirror-Image Pattern *39*
    The Mystery of the Ring of Fire *41*
    The Mystery of the Calm Coast *44*
    The Clue of the Misdirected Compass Needles *45*
    The Lost Continent *47*
    The Find on Coalsack Bluff *50*

Part II—THE UNIVERSE.........................53

3. **The Journey to the Strange World Inside the Atom**..55

   The Puzzle of the Perfect Proportions............*58*
   The Discovery of a Strange Activity.............*63*
   The Search for a Powerful Source of the New Rays...*64*
   A Clue from a Radium Gun......................*66*
   The Clue of the Atoms That Lost Weight...........*67*
   To the Inside of the Atom........................*70*
   The Bottled Lightning Experiments................*73*
   The First Exploration of the Nucleus..............*78*
   The Clue of the Upside-Down Pail of Water........*81*
   The Mystery of the Missing Charge................*82*
   Into the Nucleus...............................*83*
   The Clue of the String of Light...................*87*
   The Atom Smashers...........................*90*
   Antimatter...................................*93*
   The Ghost Particle.............................*96*
   The Clue of Too Many Hadrons..................*99*
   Into the Hadrons..............................*101*
   The Hunting of the Quark......................*103*
   What's Inside the Quarks?......................*109*

4. **The Journey to the Edge of the Universe**..........113

   The Leap to the Stars..........................*118*
   Beyond Alpha Centauri........................*124*
   The Code of the Blinking Stars..................*125*
   What Does the Universe Look Like?..............*134*
   The Discovery of Our Galaxy....................*136*
   To the Galaxies...............................*139*

      The Clue of the Speeding Galaxies............... *141*
      Seeing the Invisible Universe................... *146*
      The Clue of the Radio Waves from Space.......... *148*
      The Puzzle of the Mysterious Spectra............ *152*
      The Final Lap................................ *155*

**5. The Journey from the Beginning
to the End of Time**........................... **157**
      The Clue of the Expanding Universe.............. *163*
      The Echo of the Big Bang...................... *168*
      The Journey Through the Past.................. *170*
      Is There an End to Time?...................... *180*
      The Black Holes in Space...................... *181*

Index.. **187**

Foreword

Science fiction has taken several fantastic journeys—to the center of the earth, to the lost continent, to the strange world inside the atom, to the edge of the universe, and to the past and the future. Science has taken the same journeys, not in manned craft but by means of instruments, logic, and imagination—and these journeys have been far more fantastic.

In all tales of travel to unknown regions the explorer makes strange and wonderful discoveries. The scientific explorer has made the strangest and most wonderful of all. Floating continents, exploding stars, lands beneath the sea, ghost particles, mysterious rays, and messages in code from outer space are just a few.

The making of a scientific discovery is one of the greatest achievements of the human mind. It begins by asking a question about the unknown and then proceeds with observations, experiments, and intelligence to find an answer. The scientist is a detective solving the most baffling of all mysteries: What are things made of and what makes them work? When he has found a solution to just one part of that mystery—and there are many parts—he has made a discovery. Each of science's fantastic journeys is a tale of how scientists make their discoveries.

There is no need to be more than casually acquainted with science to come along on science's journeys. Many simple diagrams will make your trip as easy as it will be exciting. For the fullest enjoyment, read the journeys in the order in which they appear. The first two, about our earth, will accustom you to how scientists think and work. The next three are actually segments of one long journey—science's fantastic journey to find the secrets of the universe.

Preceding each of science's journeys is the story of a corresponding science-fiction journey. For me, science fiction was the gateway through which I entered the amazing world of science. It could be your gateway, too.

Part I—The Earth

1

The Journey to the Center of the Earth

There has never been an age like ours. We rocket our space ships to the dark side of the moon, and probe the mysteries at the very edge of the universe. We map mountain ranges buried beneath the sea. We journey to the strange world inside the atom, and dig out the secret of life from microscopic specks within the cell. We travel back in time to view the explosive birth of the universe. We peer into the hearts of stars. This is an age of unparalleled exploration. Where man has never gone before, we go. To where man thought he could never go, we find a way. We find a way even to the center of the earth.

To the great German scientist Professor Von Hardwigg, the clue to that way was a piece of antique parchment that fell from a yellowing old volume he was browsing through in the university library. The parchment was covered with a strange text.

It was a cryptogram, Professor Von Hardwigg discovered, a code that almost defied deciphering. But finally he broke it, and read:

Descend into the crater of Sneffels Jokul, bold traveler, and you will reach the center of the earth. As I did.

ARNE SAKNUSSEMM

The writer of the cryptogram was a famous Scandinavian scientist, and Sneffels Jokul was an inactive volcano thrusting its menacing black cone high above a glacier in northern Iceland. To test the truth of Arne Saknussemm's unbelievably wild claim, Professor Von Hardwigg, his nephew Harry, and an Icelandic guide named Hans journeyed to that glacier. They descended into the crater of Sneffels Jokul.

A long dark tunnel of basalt rock cut deep into the earth, and the three explorers followed it until they came to a great underground sea. They built a raft. They fought storms, waterspouts, and hidden reefs. Death came close to them when a titanic battle between two prehistoric monsters, an ichthyosaurus and a plesiosauras, threatened to overturn their raft. But they landed safely and found, carved into the face of a cliff close to the sea, a message left by Arne Saknussemm. *Here*, it read, *is the entrance to the tunnel that leads directly to the center of the earth*. But the entrance was blocked.

The final passageway to their goal had been sealed off by a rockfall. They would have to blast their way through. Hans rigged the dynamite. Harry set off the fuse. All three explorers then leaped on the raft and poled 20 feet into the sea. A huge explosion ripped an immense gap in the rockfall. The way to the center of the earth was clear.

But the blast had been too powerful. It had set off a tidal wave. The underground sea roared forward, sweeping through the gap the explosion had opened, plunging down the tunnel, and carrying the raft and the three men on it swiftly toward the center of the earth. But just as Professor Von Hardwigg and his companions thought they would finally reach their goal, the down-falling waters clashed with a rising stream, and suddenly they found themselves no longer going down but going up.

The up-going stream had captured their raft and was floating it back to the surface of the earth. But the stream was not water. It was red hot molten rock, lava, and already the wooden logs of the raft were beginning to burn. In seconds they would burst into flame, and a fiery death would claim every member of the Von Hardwigg expedition. But in those last seconds, the lava rose into the crater of the volcano Etna in

Stromboli, Italy, the volcano erupted, and Harry, Hans, and Professor Von Hardwigg were thrown out of the volcano's mouth to safety on the mountain's slope. THE END.

Fantastic? Of course. What you have been reading is the plot of one of the great science-fiction novels of all time, *A Journey to the Center of the Earth,* by the French author Jules Verne. It was written more than a hundred years ago, long before scientists had any accurate knowledge of what lay merely a mile beneath their feet—and it is nearly 4,000 miles to the center of the earth. Geologists have since made that journey to the center of the earth—not actually, but by means of their instruments—and what they have discovered is an underground world far more fantastic than even the brilliant imagination of Jules Verne could conceive. What they have done is taken a journey that matches its fictional counterpart in adventure and suspense. And, strangely enough, the story of their true journey to the center of the earth begins, like Verne's, with a cryptogram that only one man could decipher.

The Puzzle of the Double Tremor

The kind of cryptogram that the Yugoslavian geologist Andrija Mohorovicic puzzled over on October 8, 1909, looked like this.

It was a message written by an earthquake.

As far back as 1846 the English geologist Robert Mallet had discovered that earthquakes send out shock waves that can be felt thousands of miles away. You can create waves when you toss a pebble into a pool; you can see the waves spreading out on the water. But no pebble can produce a wave in the rock-hard body of the earth. To send waves vibrating through continents requires a gigantic force. The smallest of our major earthquakes releases more energy than the atom bomb that destroyed Hiroshima. The earthquake that killed thirty thousand people in Assam, India, on August 15, 1950, had the energy of more than one hundred thousand atom bombs.

To capture shock waves on paper so they could be studied, scientists of Mohorovicic's time had invented an instrument called a seismograph.

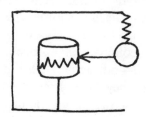

It is a simple device. A sort of hard, sharp pencil point, a stylus, is attached to the side of a suspended heavy pendulum. The stylus rests against a sheet of paper wound about a rotating cylinder. The pendulum keeps the stylus in a fixed position at all times, even during earthquakes. When the earth is at rest, the stylus inscribes a straight line as the cylinder turns. But when the earth trembles, so does the cylinder. The turning cylinder swings and sways with the motion of each shock wave, and the straight line breaks into a pattern of twists and curls. The stylus writes out a message from an earthquake in a strange language of squiggles.

It was a message of this sort that perplexed Mohorovicic on that October day in 1909, a day when the first major breakthrough in science's knowledge of the interior of our planet was to occur.

Seismologists, the scientists who specialize in earthquakes, had learned how to translate the language of the squiggles. They could read the warps, curves, and bends and tell with astonishing accuracy the intensity of the shock, just how far the shock had traveled, and the moment when the earthquake had struck. By coordinating the messages from two seismographs in different locations, they could even pinpoint the exact location from which the waves had begun to spread, the epicenter of the earthquake. The language of the squiggles told them all this and a good deal more—and Andrija Mohorovicic, a professional seismologist, could read the language easily. It wasn't how to read the message from the earthquake that puzzled him; it was the message itself.

In Mohorovicic's translation the message (actually the co-ordinated results from two seismographic stations) read:

Epicenter: the Kulpa Valley, Croatia. A double tremor: a strong earthquake, followed shortly thereafter by a weak one. The message gave the exact times when the earthquakes had struck.

So far nothing out of the ordinary. A strong set of earthquake waves had shown up on the seismographs. A short time later, a weak set had shown up. Both had originated from the same epicenter. This simply meant that the earthquake was a double tremor, or double shock, as familiar to seismologists as thunder and lightning on a summer's day is to weathermen. Mohorovicic would have thought no more of it had he not received at the same time another message in the language of the squiggles. This message came from seismographic stations more than a hundred miles away. He translated it. The message read:

Epicenter: the Kulpa Valley, Croatia. A double tremor: a weak *earthquake, followed shortly thereafter by a* strong *one.* It gave the exact times when the earthquakes struck.

The times in each message were identical. The epicenters were identical. There was no doubt that each message was describing the same double tremor. But one message read: A *strong* earthquake followed by a weak one; and the other message read: A *weak* earthquake followed by a *strong* one.

The second message was the reverse of the first, yet it had been sent out by the earthquakes at the same time that the first had. It was as if one message had gone out with two meanings, one meaning the opposite of the other. That was impossible, yet it had happened. The familiar message from a double tremor had become a puzzle.

To Mohorovicic that could mean only one thing: He was not translating the message accurately. Seismologists thought they had completely broken the code of the strange language of earthquakes, but, Mohorovicic concluded, they must be wrong. The puzzle of the double tremor was a cryptogram that had not yet been deciphered. Mohorovicic decided to decipher it. But he needed a clue. He set about to find one.

He collected messages, seismograms, from the Kulpa Valley double tremor from all the seismographic observatories in Europe. Some of these seismograms showed a strong tremor followed by a weak one. He called these seismograms the strong-weak messages. Other seismograms showed a weak tremor followed by a strong one, and he called these the weak-strong messages. He placed all the strong-weak messages in one pile and all the weak-strong messages in another.

Then he examined the messages in the strong-weak pile to find out if they had anything in common other than what he already knew. They did. Every one of the strong-weak messages came from observatories *inside* a 100-mile radius of the tremors' epicenter. What about the messages in the weak-strong pile? Every one of those messages came from observatories *outside* that 100-mile radius. Mohorovicic had found his clue. It was this:

The strong-weak messages flip-flopped into weak-strong messages at a definite distance from the epicenter.

The solution to the cryptogram of the double tremor now hinged on the answer to why the flip-flop occurred at a definite distance from the epicenter. Mohorovicic came up with a brilliant explanation.

The interior of the earth, he reasoned, must be divided into two different regions. Earthquake waves are slowed down through one, as if they were racing on a slow track. Here, in a highly simplified diagram of a cross-section of the earth, is that slow track in the vicinity of the Kulpa Valley.

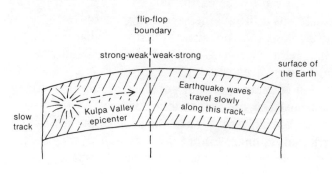

The other region speeds up earthquake waves like a fast track.

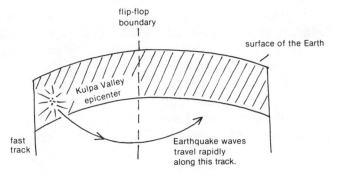

Now, Mohorovicic continued, let us suppose that there is one tremor at the Kulpa Valley epicenter, not two. The waves race along two different tracks. One set of waves, *A* in the following diagram, follows the slow track. The other set of waves, *B* in the following diagram, follows the fast track. Which set, *A* or *B*, will arrive first at a seismographic observatory outside the flip-flop boundary?

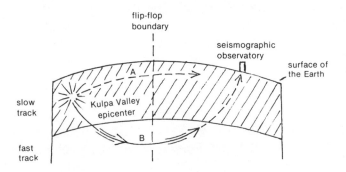

The *B* waves, even though they travel a longer distance than the *A* waves, pick up so much speed on the fast track that they beat the *A* waves to the finish line. At the finish line the *B* waves, having traveled that longer distance, are weaker than the *A* waves that come in second.

Assuming that there was just one tremor during the Kulpa Valley earthquake, Mohorovicic went on, then two sets of waves would have arrived at observatories outside the flip-flop boundary. The first set of waves would have been weak; the second set, strong. That was exactly what the squiggles from those observatories had shown.

It was also clear to Mohorovicic why a boundary seperated strong-weak messages from the weak-strong messages. Up to a definite distance from the epicenter,

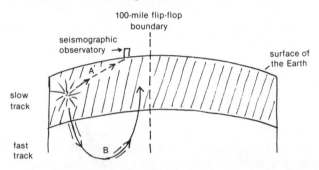

the speed that *B* picked up on the fast track did not compensate for the greater distance the waves had to travel, including two stretches on the slow track. The *A* waves arrived at the finish line first. Since they traveled a shorter distance than the *B* waves, they arrived stronger than the *B* waves. The two sets of waves produced by the single tremor arrived at observatories inside the flip-flop boundary at different times; the strong waves first, then the weak waves—exactly as recorded in the messages from the earthquake.

Andrija Mohorovicic had solved the puzzle of the double tremor. By doing so, he had discovered two separate layers within the earth. The slow track, the upper layer, he called earth's outer crust. The fast track, the lower layer, he called earth's inner crust. From the speed of the earthquake waves and the distances from the epicenter to observatories, he was able to calculate the depth of the inner crust. It descended about 1,800 miles below the surface of the earth. The first long lap had been taken to the center of the earth, still about 2,200 miles farther away.

What were the outer and inner crusts made of? Mohorovicic thought he could tell. Earthquake waves travel through different substances at different speeds. Knowing the speed of the waves, he could identify the substances. The outer crust, he found, is made up of the same kind of rocks, granite and basalt, that appear on the surface of the earth. But he was astonished to discover that the speed of the earthquake waves passing through the inner crust corresponded to no known substance. The inner crust of the earth was composed of something strange.

To find out more about this mysterious substance, geologists needed samples—samples from a region where no man had ever gone before. How were they to get them?

The Clue of the Stone Icicles

What conceivable ways were there to journey through the outer crust of the earth? Through caves? Some caves wind for hundreds of miles underground, and few had been thoroughly explored. Was it possible that some still undiscovered tunnel might lead a spelunker, a scientist who explores caves, directly into the earth's inner crust?

To answer that question, geologists asked another. According to Mohorovicic's calculations, the inner crust begins 20 to 40 miles below the earth's surface; are there caves that penetrate that far down? These geologists found the answer without stirring from their laboratories. All they needed was the clue of the stone icicles.

Iciclelike formations hang from the ceilings of caves. These stone icicles begin to form when drops of water seep through a bed of limestone above a cave. (Limestone beds are the compressed remains of small animals dead millions of years.) The water dissolves the limestone, and the drops ooze out onto the ceiling of the cave. There the water evaporates, leaving behind a tiny limestone deposit that sticks to the ceiling. When another drop evaporates on the same spot, the limestone deposit grows. As millions and millions of these drops evaporate year after year, the deposits elongate into the shape of icicles.

Because of the presence of stone icicles (called stalactites) in caves, geologists concluded that wherever there are caves there must have been limestone and water in the past. Caves were formed, these geologists reasoned, when streams of water—not drops which evaporate easily—washed through limestone beds, and carried away the dissolved limestone, leaving cavities in the earth. A residue of limestone formed the ceilings of some cavities, and the ceilings and cavities were enclosed by basalt and granite rock, which water cannot dissolve.

The way in which caves were formed held the answer to whether caves could exist 20 to 40 miles below the earth's surface. Flowing water is necessary to make a cave, and flowing water is seldom found farther down than a mile and a half. Caves could not be formed 20 to 40 miles down. Caves occur on the earth's surface or just below it. If a spelunker attempted to find his way to the earth's inner crust by following the dark passages of caves, he would soon come to a dead end of solid rock. From the clue of the stone icicles, geologists determined that a cave cannot be a natural passageway through the earth's outer crust.

Project Mohole

By the beginning of the 1960s geologists had still found no way to bring up samples from the inner crust. Mining technology had taken gigantic strides forward, and a mining shaft in India had penetrated the earth for a distance of about two miles. But building a mining shaft to transport men and equipment 20 to 40 miles into the earth presented more problems than sending a manned spaceship to Jupiter. Besides, scientists were certain, no human being could live in the intense heat inside our globe.

Temperatures rise about 16 degrees Fahrenheit for every 1,000 feet of depth. In one of South Africa's gold mines, the Robinson Deep, which goes down about 1¼ miles, a powerful air-conditioning plant had to be installed in order to prevent the miners from literally being roasted alive. At a depth of only 1½ miles, the temperature of the rock reaches the boiling point of water. (When water seeping from the earth's surface reaches this depth, it begins to boil and blows out of small cracks in the earth's outer crust in the form of geysers, like the ones that continually astonish visitors at Yellowstone National Park.) The boiling point of water is 212 degrees Fahrenheit, and 120 degrees Fahrenheit may be the upper limit of man's ability to tolerate heat. Within the inner crust temperatures soar so high—in excess of 3,300 degrees Fahrenheit—that rocks begin to melt.

No team of geologists could be sent to take samples from the inner crust. But where man cannot go, man-made machines often can. Geologists thought they could bring up samples of the mysterious substance in the inner crust by using a boring rig. The oil industry had made deep drillings with this kind of machinery for years. The equipment, the know-how, and the experienced personnel were ready. There was only one obstacle. No boring rig had penetrated the earth for a distance greater than 5 miles, and the shortest distance to the inner crust, according to Mohorovicic's calculations, is about 20 miles.

Mohorovicic had arrived at this figure by calculating the points at which earthquake waves began to speed up in the area around the Kulpa Valley. These speedup points marked the existence of a boundary between the outer and the inner crust. That boundary is called the Mohorovicic discontinuity, or simply the Moho. Other seismologists calculated speedup points throughout the world and found that the distance to the Moho varies. The average distance *is* 20 miles, but they discovered one site where the thickness of the outer crust is only 10 miles deep. But 10 miles was twice as far as any boring rig had ever drilled.

The search began for a location where the earth's crust is thinner than five miles. Geologists at the Lamont-Doherty Geological Observatory, near New York City, studied the speedup points of earthquake messages from all over the globe and found the place. Off the coast of Puerto Rico the distance to the Moho is only two to three miles. But geologists were not certain they could drill there. The earth's surface in this area is blanketed by the Atlantic Ocean, and before the cutting edge of the drilling rig could bite into the earth, it would have to descend through 4 miles of water. It had never been done before. Could it be done?

In a test conducted by a Lamont-Doherty research team off La Jolla, California, in 1961, a cutting edge was lowered 3,000 feet to the floor of the Pacific. The drill then penetrated 1,035 feet into the earth's crust under the ocean. This was an encouraging result, and geologists rushed to improve their deep-water drilling equipment and techniques. Within a year a highly trained crew handling superior equipment sunk a cutting edge through more than 2 miles of ocean off Guadalupe, Mexico, and drilled into the earth for a distance of 601 feet. This proved the feasibility of attempting to drill a hole under 4 miles of ocean to penetrate the thin outer crust of the earth off the coast of Puerto Rico. The American National Science Foundation approved a project to do so. Since the hole would go through the Moho, the project was called Project Mohole.

The Mohole drill was designed to take a continuous sample, or core, of the earth through which it passed. As the drill descended, it would first pick up material from the ocean floor and then from the earth's outer crust. Finally, after breaking through the Moho, the drill would collect samples of the mysterious substance in the inner crust. But Project Mohole ran out of funds and was shelved. Another way had to be found to break through the Moho.

The Clue of the Undersea Map

The clue to the passageway through the Moho came from a map so strange that the scientists who first saw it refused to believe it.

Marie Tharp, a draftswoman working for the Lamont-Doherty Geological Observatory, made the map. In the early 1950s, the observatory had sent out a fleet of laboratory ships to survey the bottom of the Atlantic Ocean with sonar, sound echoes like those that enable bats to "see" in the dark. The laboratory ships had obtained thousands of echo patterns from the entire bed of the Atlantic. It was Marie Tharp's job to translate those patterns into a three-dimensional map, a model of the ocean's bottom. When she showed her completed work to the heads of the project, they told her that what she had constructed was just too fantastic; surely she had made a mistake. But when the sonar data and Marie Tharp's interpretation of that data were checked, the project heads had to concede that her map was accurate.

Before Tharp constructed that map, scientists could only guess what the undersea terrain looked like. To most of them, the ocean floor was a flat plain of ooze, broken only by an occasional mountain or deep chasm. No man had ever journeyed to the bottom of the sea to chart the unknown regions beneath miles of frigid, sunless waters exerting pressures that in an instant could crush bars of some metals into powder. In the comfort of her workshop Marie Tharp had charted the floor of the Atlantic Ocean and made one of the greatest discoveries in the history of exploration. She found an amazing new land underneath the sea.

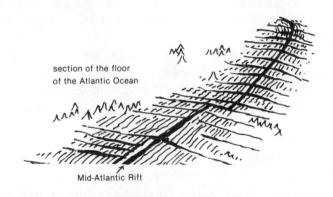

section of the floor of the Atlantic Ocean

Mid-Atlantic Rift

On both sides of the Atlantic, the continents slope sharply into the ocean. Stretching out from the continental slopes are the plains that many geologists had predicted, long flat stretches covered with soft sediment washed from the continents by rain and rivers. Toward the center of the ocean bottom the plains give way to small lumpy hills. At the center of the ocean bottom, running north and south for the entire length of the ocean, is one of the mightiest mountain chains on earth. This is the

Mid-Atlantic Ridge, actually two parallel mountains separated by a valley 8 to 30 miles wide.

It is no ordinary valley. Midway between the continents that flank the Atlantic, and from one end of the ocean to the other, the earth's crust had cracked open. That crack is the valley, a deep cleft called the Mid-Atlantic Rift, splitting the floor of the Atlantic in half.

J. Tuzo Wilson, a Canadian geologist, was trying to explore the inner crust, and he thought that this almost unbelievable crack in the earth might be a gateway through the Moho. Did the Mid-Atlantic Rift go down deep enough to cut into the inner crust? he asked. No geologist could supply the answer.

The Clue of the Undersea Volcanoes

Wilson knew that since 1957 undersea volcanic activity had been reported in the Atlantic. In that year an underground volcano erupted off the island of Fayal in the Azores. In 1961 a volcanic explosion shook the waters of the South Atlantic at Tristan de Cunha. A similar explosion hit at Askja in Iceland. In 1963 a new volcanic island, Surtsey, was blown into the Atlantic off the coast of Iceland.

When Wilson pinpointed the eruptions on Marie Tharp's undersea map, he found each new volcano was located on the Mid-Atlantic Rift. The crack in the earth's crust was a birthplace of volcanoes. This was the clue Wilson needed.

He knew that volcanic lava is formed when the temperature inside the earth rises high enough to melt rocks. This occurs at a depth of about 30 miles. To provide lava for volcanoes, the crack in the earth should go down at least that far. Under the Atlantic the distance to the Moho is only 2 to 3 miles. If the crack did go down 30 miles, it was a gateway to the Moho. Did it?

Wilson looked for proof. If the crack did go down to the region inside the earth where lava forms, then lava should well up from the crack and spill out on the ocean floor. The lava would then cool and solidify into a specific kind of rock called igneous rock. If Wilson could find igneous rock on both sides of the crack, it would be proof that the Mid-Atlantic Rift cut into the inner crust. Earthquake-wave studies had been made of the regions adjacent to the Rift. These studies showed that the regions on both sides of the Rift were composed entirely of igneous rock. Wilson had found a gateway through the Moho.

It was an unusual gateway. Wilson did not have to pass through it and take out samples of the inner crust; the gateway brought out samples to him: the igneous rocks on both sides of the Rift. Further research showed that these samples covered the ocean floor from the Mid-Atlantic Rift to the slopes of the continents.

But two different kinds of analysis of these samples presented a puzzle. Deep-sea drills bit out chunks of these samples. Chemical analysis of them showed the igneous rock was composed mainly of iron and magnesium. Geologists agreed that at the temperatures within the inner crust, rock of this type would melt. They concluded that the inner crust was composed of liquid rock rich in iron and magnesium. Earthquake analysis, which had reached a high degree of perfection, confirmed the iron-magnesium content of the rock in the inner crust but added a disconcerting message: The rock was solid. The puzzle geologists faced was this: How could a rock be solid and liquid at the same time?

It was an incredible contradiction. But Wilson found a way to explain it.

The Clue of the Hot Ice

In the early 1960s a team of researchers at Harvard University had subjected a block of ice to enormous pressure: 600,000 pounds on every square inch. Then they had heated the ice to well over the boiling point of water (ice melts at 32 degrees Fahrenheit; the boiling point of water is 190 degrees higher), but the ice had not melted. Instead, it had become flexible like gum or a plastic material, and force exerted on it could make it flow like a liquid. The hot ice under extreme pressure had become solid and liquid at the same time.

What had happened to the hot ice, Wilson concluded, must be happening to the hot rock of the inner crust. The miles-high layer of granite and basalt rock of the outer crust exerts enormous pressure on the iron-magnesium rock of the inner crust. Under this pressure the hot inner-crust rock does not melt but becomes a flexible solid that flows like a liquid.

With this conclusion Wilson solved the mystery of the strange substance Mohorovicic had found in the inner crust. But the solution of this mystery led to another. What effect did this solid-liquid material of the inner crust have on the outer crust, the thin layer of earth on which we live? Surely, Wilson thought, in that massive sampling of matter from the inner crust that spread across the ocean floor there must be a clue to enable him to find out what was happening down there. In the spring of 1963 Wilson received a startling report that contained that clue.

The Clue of the Flip-flopping Compass Needles

The report came from the Canadian geologist L. W. Morley. Morley began his investigations, the report read, with one of the most commonplace of scientific facts: Magnetized iron points north. That is the principle of the compass, and it has been taken for granted by civilized man for hundreds of years. Iron-bearing rocks on the earth's surface, Morley went on, contain thousands of minute particles of iron, tiny compass needles pointing north. Since the rocks of the floor of the Atlantic Ocean are rich in iron, Morley expected that they, too, would contain these compass needles. They did. He also expected that across the entire ocean floor all the compass needles would point north. They did not.

Rolling out from each side of the Mid-Atlantic Rift, Morley discovered, the rock carried a strange pattern of compass readings. For several miles the needles pointed north, then flipped to south for a number of miles, then flopped back to north for several miles more, then flipped to south again ... and continued to flip-flop across the ocean floor from the Mid-Atlantic Rift to the continents. Here is a simplified diagram showing the flip-flops in compass readings that Morley found in the iron needles within the igneous rocks as he followed one section of the ocean floor to the west of the Rift. (He found the identical pattern to the east of the Rift.)

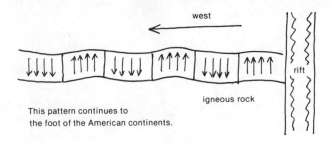

This pattern continues to the foot of the American continents.

J. Tuzo Wilson realized, as he finished reading Morley's report, that the flip-flop pattern of compass readings was formed on rocks that had originated in the earth's inner crust. The pattern had to be a clue to what was happening down there. Wilson set out to discover how that pattern was formed.

He began by asking: How did each segment of the ocean floor get its present compass reading?

To find the answer, he first reviewed what was known about compass needles. All compass needles should point north because the earth is an enormous magnet, and all compass needles must point to a magnet's north pole. The earth's magnetic north pole is not at the earth's geographic north pole, but the two poles are close to each other. But in the remote past, according to scientific findings made in the early 1960s, the earth's north magnetic pole was frequently located close to the earth's *south* geographic pole.

This change is described by geologists as a reversal of the earth's magnetic field. Every time the earth reversed its magnetic field, the north magnetic pole changed its geographic location: from north to south, then from south to north, then from north to south, and so on. Over the last 76 million years, at irregular periods ranging from 30,000 years to 2 million years, the earth has reversed its magnetic field 171 times. And every time the field reversed, the direction of natural compass needles reversed as well.

So, it became clear to Wilson, each segment of the floor of the Atlantic Ocean must have taken on its present compass reading sometime in the past. For example, a segment now near the North American continent

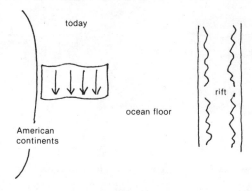

had been formed tens of millions of years ago when lava welled out of the Mid-Atlantic Rift. As the molten rock solidified on the ocean floor, the tiny compass needles in it were magnetized by the earth's magnetic field to point in the direction of the earth's magnetic pole at that time. When the rock was completely solidified, the direction of the compass needles in it was frozen permanently. Wilson had found out how the flip-flop pattern had been formed. But his discovery created a new puzzle. How could solid rock travel all the way from the Rift

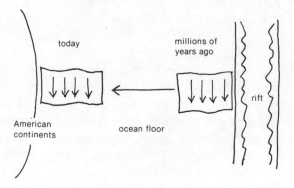

The Journey to the Center of the Earth

to the edge of the continent? He reasoned:

Something had to carry the rock across the ocean floor. But the rock was firmly embedded in the earth's outer crust. Something, therefore, had to carry the earth's outer crust across the ocean floor. That something, Wilson knew, was the solid-liquid rock that made up the inner crust. His calculations showed that as a solid the inner-crust rock had the strength to hold up the outer crust, and as a liquid it had the mobility to move it.

But, Wilson asked, what made the solid-liquid rock of the inner crust move the outer crust to the west on one side of the Rift and to the east on the other side?

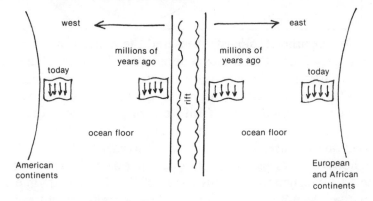

Springing off from his knowledge of what happens when liquids are heated, Wilson made a bold imaginative leap. He knew that heat sets up special types of currents in liquids called convection currents. What makes these currents special is that they always come in pairs.

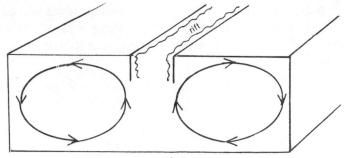

Convection currents

Convection currents in the inner crust, Wilson calculated, are thousands of miles wide. They carry the earth's outer crust along with them to the west and to the east of the Rift, just as if the outer crust were on huge conveyor belts.

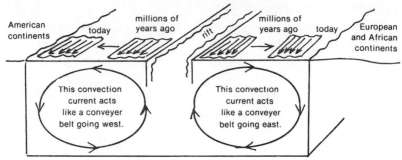

When other geologists measured the motion of the ocean floor, they found that it moves about an inch a year. That is the rate of speed of the convection currents. It took millions of years for the convection currents to carry segments of the ocean floor from the Mid-Atlantic Rift to their present positions.

Wilson reconstructed the history of the Atlantic Ocean, based on the existence of convection currents in the inner crust. Millions of years ago the ocean floor split open, and soon after, the first magnetized igneous rock solidified. At that time the Atlantic was not much wider than the Rift itself—50 miles at the most. Then the giant conveyor belts inside the inner crust moved the ocean floor away from each side of the Rift. New igneous rock welled out of the Rift to replace the floor that had been moved. The ocean floor spread about an inch a year to its present width.

Wilson's idea of ocean-floor spreading raised still another question: What happened to the spreading ocean floor when it came to the locations where the convection currents tipped down?

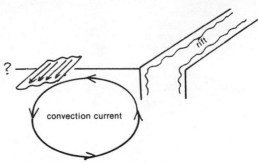

Wilson worked out the answer. The convection currents carried the igneous rock back into the inner crust. There it was converted by intense heat and pressure to a solid-liquid rock and transported by the convection currents back to the crack in the earth to start another cycle.

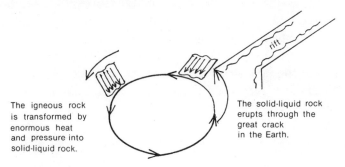

The igneous rock is transformed by enormous heat and pressure into solid-liquid rock.

The solid-liquid rock erupts through the great crack in the Earth.

This process of birth, destruction, and renewal of the ocean floor, Wilson concluded, has been going on for millions of years, and is still going on. It will continue, he predicted, as long as the earth is able to create its own interior heat to keep the circulatory convection currents flowing.

The clue of the flip-flopping compass needles had led K. Tuzo Wilson through a series of questions and answers to an understanding of what was happening down there in the inner crust. He had explored 1,800 miles into the earth.

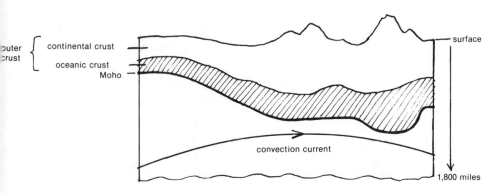

Cross section of a segment of the Earth, not drawn to scale. The outer crust is actually very thin compared to the size of the Earth. If the Earth were an apple, the outer crust would be the skin.

What lay beneath? To find the answer, and continue their journey to the center of the earth, geologists had to solve a baffling mystery.

The Mystery of the Shadow Zones

In the spring of 1954 the United States exploded four megaton-sized hydrogen bombs at Eniwetok in the Pacific. When the seismographic data from this man-made earthquake were mapped, puzzling areas appeared on the side of the globe opposite the epicenter. They were shadow zones, in which no earthquake waves were detected, even though the waves had been detected everywhere else.

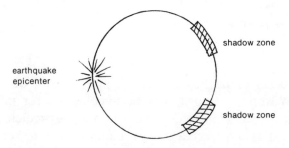

More puzzling was the presence between the shadow zones of a region where seismograms indicated an extremely high concentration of earthquake waves.

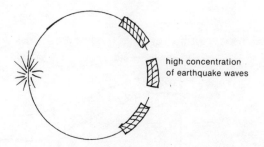

The pattern of a region of highly concentrated earthquake waves surrounded by shadow zones did not appear only as a result of the Einwetok explosion. No matter where a major earthquake hit, seismologists would find the same pattern on the side of the globe opposite the epicenter.

Geologists came to two conclusions: The first was that something inside the globe is blocking earthquake waves and preventing them from getting through to the other side. The second was that something else inside the globe is focusing earthquake waves and concentrating them when they get to the other side.

Those somethings, geologists knew, are not located in the upper or inner crusts, because earthquake waves easily pass through these regions without evidence of blockage or concentration. Those somethings had to be located below the inner crust, in the earth's core. Finding out what those somethings are would enable geologists to map the remaining unknown territory inside our planet and reach the end of their journey to the center of the earth.

Teams of geologists all over the world, operating independently and pooling their results, tackled the problem. They knew that seismographic squiggles are caused by two kinds of earthquake waves: P waves that result when the earthquake, like a giant hammer, hits the earth's crust head on,

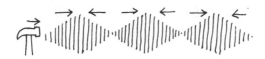

and S waves that result when the earthquake strikes at an angle.

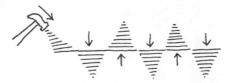

All earthquakes strike in both directions at once. The P waves and S waves show up as different kinds of squiggles on the seismograms. A modern seismograph, in which the stylus is replaced by a beam of light acting on a photographic plate, captures the P- and S-wave squiggles with extreme accuracy.

Geologists seeking to solve the mystery of the shadow zones analyzed the P- and S-wave squiggles from many earthquakes and found a clue. Between the epicenter and the fringe of the shadow zones, both P and S waves came through the earth. But in the region between the shadow zones where there was a high concentration of earthquake waves, only the P waves came through.

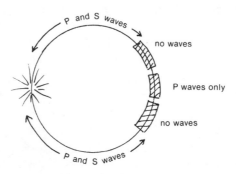

It was known that S waves are blocked by liquids. The core of the earth, geologists concluded, is liquid and deflects S waves.

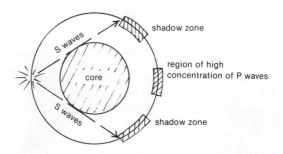

This explains why there are no S waves in the shadow zone.

But P waves, geologists knew, could pass through both solids and liquids, and yet no P waves could get through to the shadow zones. That could only mean one thing, the investigating scientists concluded: material inside the core focuses P waves much as a lens focuses light waves.

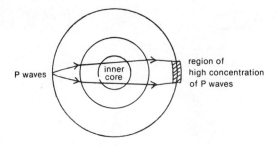

This explains why there are no P waves in the shadow zones.

The mystery of the shadow zones had been solved. From the extent of the shadow zones and the zone of high P-wave concentration, geologists calculated that the outer core extended into the earth to a depth of 3,150 miles. Analysis of the speed of the earthquake waves passing through it showed that it is composed of molten metal, mostly iron, nickel, and chromium. Temperatures are many times higher than those of the inner crust.

The inner core, earthquake-wave analysis showed, is a dense metallic solid. It does not melt, even though the temperatures in the inner core are the highest on our planet. It cannot melt because enormous pressure is exerted upon it—pressure provided by the weight of most of our planet above it. Earthquake-wave analysis identified the material of the inner core as 10 percent chromium and 90 percent iron. But even closer to the actual center of the earth, the chromium disappears. Geologists, at the end of their 4,000-mile journey through the earth's interior, had found the center of the earth to be a super hot ball of iron several miles in diameter.

In their descent to the earth's core, geologists left several major mysteries unsolved. What are the reasons for earthquakes, and for volcanoes not located along the Atlantic Rift? How were mountains, other than volcanic peaks, born? Are there rifts in other oceans besides the Atlantic? If so, are other ocean floors in motion? What effect does the moving ocean floor have on continents? Why are continents located where they are, and how did they get their shapes?

The answers would come, as answers to the unsolved mysteries of one scientific quest often do, from another quest toward a new goal. In pursuit of that goal, geologists would make one of the most astonishing discoveries of our time: The continents are adrift on the face of the globe. And the new goal? To find a lost continent.

2

The Journey to the Lost Continent

The first expedition to the lost continent was headed by the renowned British scientist Dr. Maracot. He was assisted by two Americans, Cyrus Headley of the Zoological Institute of Cambridge, Massachusetts, and Bill Scanlon, an engineer on leave from the Merribank Works, Philadelphia. Their destination was a location in the Atlantic Ocean southwest of the Canary Islands where Dr. Maracot had previously discovered a 5-mile-deep pit in the ocean floor, the Maracot Deep.

When Dr. Maracot's laboratory ship *Stratford* arrived over the Maracot Deep, the three members of the expedition entered a steel sphere equipped with glass windows for observation. The sphere was lowered into the Atlantic from the ship by means of a steel chain, a hawser, to which several air lines had been attached. The immediate goal of the expedition was a ledge about ⅓ mile down overlooking the Deep. As the sphere rested on the ledge, it was attacked by a gigantic lobsterlike creature that took hold of the hawser, swung the sphere directly over the Deep, and cut the hawser with its monstrous claws. The sphere plunged to the bottom of the Deep.

The air lines had been severed, and the three explorers were about to die when Headley saw a human face looking in through one of the windows. Believing that rescue was near, Dr. Maracot opened the entrance hatch. The pressure of the gases inside the sphere held the water from rising above chest level. Within seconds two men wearing transparent watertight suits over their heads and bodies entered the sphere. The suits were equipped with a device for supplying fresh air. The two men from the sea quickly gave suits like those they were wearing to Dr. Maracot and his assistants.

The explorers were led from the sphere to a huge building in a strange underwater city, where they met a people speaking a language that even the scholarly Dr. Maracot could not identify. These people could communicate, though, by throwing images of their thoughts on a screen. From these images the explorers learned that the city under the sea was the capital of the lost continent of Atlantis. Thousands of years before, a sudden cataclysm had torn the continent from the face of the earth and cast it into the ocean depths. Some Atlanteans had prepared for the catastrophe and were able to survive on the ocean floor. Their descendants had continued to live there ever since.

Dr. Maracot, in his exploration of undersea Atlantis, entered a dark, sinister temple shunned by Atlanteans. He encountered the Lord of the Dark Face, the long-lived evil genius who had produced the cataclysm that had sunk Atlantis. The Lord of the Dark Face announced that the Atlanteans had survived too long and he would soon wipe out the entire race. Dr. Maracot engaged the Lord of the Dark Face in a duel to the death fought with will power. The power of Dr. Maracot's will came from the force of good; the power of the Lord of the Dark Face's will came from the power of evil. Dr. Maracot won, the Lord of the Dark Face died. The people of Atlantis were saved.

In gratitude the Atlanteans presented Dr. Maracot, Headley, and Scanlon with large glasslike balls filled with a light gas. The explorers attached themselves to the balls, rose to the surface of the Atlantic, and were rescued. THE END.

You have just read a synopsis of *The Maracot Deep,* one of the two great science-fiction novels written by Sir Arthur Conan Doyle, best known as the creator of Sherlock Holmes. (Doyle's other science-fiction novel is *The Lost World,* the original yarn about prehistoric dinosaurs still alive in modern times.) In 1928, when Doyle wrote *The Maracot Deep,* the idea of a lost continent was regarded as fantastic. But since then geologists have found a lost continent—not Atlantis, but a supercontinent that existed on earth until about 200 million years ago.

In search of that lost continent geologists followed a trail that began at the bottom of the ocean, went on to drifting volcanic islands, continued to a mysterious ring of fire encircling the Pacific, and ended on a mountain close to the South

Pole. How they discovered that trail is a thrilling story of scientific detection stranger even than the adventures of the Maracot expedition. The story begins, as most detective stories and all major scientific investigations do, with a mystery.

The Mystery of the Missing Land Bridges

As the 1920s ended, geologists continued to mull over a decades-old puzzle. It centered on a big-leafed flowerless plant, *Glossopteris,* which had lived about 350 million years ago. Paleontologists, scientists who study fossils—the remains of plants and animals of ancient eras—had found that *Glossopteris* had flourished in South America, South Africa, Australia, India, and Antarctica. Modern plants of the same kind as *Glossopteris* first appear in one place and then spread to others by means of spores (small bodies with which some plants make copies of themselves). The spores are carried by wind, insects, or animals over short distances of land. But the continents on which *Glossopteris* had appeared are separated by hundreds to thousands of miles of water.

What puzzled paleontologists was this: How did *Glossopteris* spread from one of these continents to all the others when its spores could not be carried over water?

To some geologists there was only one possible answer: *Glossopteris* spores had been carried over land bridges that had since disappeared under the sea.

[Diagram: Outlines of South America, Africa, India, Australia, and Antarctica with dotted lines connecting them. Caption: "Dotted lines indicate possible sites of ancient land bridges."]

These land bridges had linked the five southern continents into a single continent which these geologists called Gondwanaland (after a region in India rich in *Glossopteris* fossils).

There could be no question that Gondwanaland had once existed if the sunken land bridges could be found. But in the 1920s geologists had no way of exploring vast expanses of the ocean floor. It was not until the late 1950s that geologists were able to journey to the bottom of the sea in search for the sunken land bridges. They made that journey not with undersea craft but with an instrument that takes advantage of a remarkable characteristic of the ocean.

That characteristic is the capacity to transmit sound over great distances—far greater distances than sound can be transmitted by air. The explosion of a one-pound block of dynamite in air, for example, can be heard for half a mile; in mid-ocean depths, the explosion can be heard for thousands of miles. Sound waves from a ship can easily reach the ocean's greatest depth (about 6.7 miles), and the echoes can just as easily return to the surface. It is with sound waves and their echoes that geologists reach the ocean floor.

The technique for doing this is called echo sounding. The principle is simple: A surface ship sends sound pulses—pings—downward, and measures the time it takes a ping to reach the bottom and echo back. Since the speed of the sound is known, the distance to the bottom can be calculated. A ship equipped with an echo-sounding instrument (sonar) can make an accurate continuous record of the topography—the ups and downs—of the ocean floor over which the ship passes.

Echo soundings taken by many ships over thousands of miles of the southern oceans revealed no land bridges. "Land bridges could not disappear without a trace," commented Don Tarling, an eminent British geologist, "[and since] all our knowledge of the ocean has still not revealed even the slightest trace of such sunken continental links . . . it *must* be concluded that these extensive land bridges never existed [Italics his]."

But *Glossopteris* could only have spread over land. If the land bridges had not sunk beneath the sea, what had happened to them? For years geologists could shed no light on the mystery. Then, in the early 1960s geologists engaged in a routine job of determining the age of volcanic islands came up with an exciting clue. To follow up on that clue, geologists in search of the land bridges of the lost continent of Gondwanaland journeyed from the bottom of the sea to the islands of the Azores.

The Clue of the Aging Volcanic Islands

This diagram represents the chain of nine islands that form the Azores, an archipelago in the Atlantic some 800 miles off the coast of Portugal.

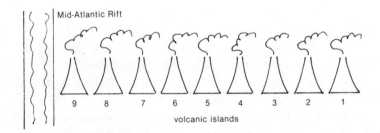

To determine the age of these islands, geologists read a clock that is built into many rocks. That clock is actually the element potassium-40. Potassium-40 is radioactive, which means it radiates, or gives off, submicroscopic particles. When this happens, some of the potassium-40 changes into another element, argon-40. Potassium-40 changes to argon-40 at a fixed speed, just as the hands of a clock change position at a fixed speed. The potassium-40 clock begins to tick (scientists say it begins to decay into argon-40) just as soon as the rock has formed. By measuring the amounts of potassium-40 and argon-40, geologists can calculate how long the clock has been ticking. In this way they can tell the age of the rock.

When geologists read the age of the oldest rocks on each of the nine islands of the Azores, they found that

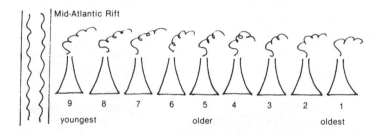

the island nearest the Mid-Atlantic Rift was the youngest and that the islands grew progressively older the farther they were from the Rift. This was the clue that brought the geologists who were searching for the lost land bridges to the Azores. (They made the journey without leaving their laboratories; they merely read the reports of the geological expeditions that had made the clock readings.) From this clue, these geologists concluded, something almost unbelievable had occurred in the Azores, which could help explain the mystery of the missing land bridges. Here is how they reached that conclusion:

They knew from the work of J. Tuzo Wilson and others that volcanoes are born along the Mid-Atlantic Rift. They also knew that Wilson had discovered giant conveyor belts inside the earth (convection currents) that carried the ocean floor along with them. Putting these two facts together, these scientists explained why the oldest islands are the farthest away from the Mid-Atlantic Rift. Volcanic island 1 was born on the Rift

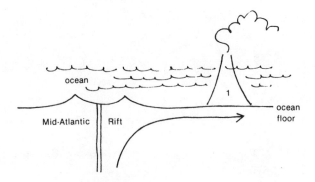

and then carried by convection currents to the east. Then volcanic island 2 was born on the Rift

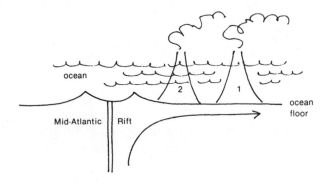

and was carried along after island 1. The rest of the islands followed. The almost unbelievable conclusion was: The nine islands of the Azores moved along the face of the earth.

This discovery revived an idea dealing with the lost land bridges proposed by the German geologist Alfred Wegener as early as 1912. According to Wegener the present-day continents that formerly made up Gondwanaland were not linked by land bridges, but were fitted to each other like pieces of a jigsaw puzzle. Then the pieces separated and drifted away from one another. As evidence, he pointed to the matching coastlines of South America and Africa.

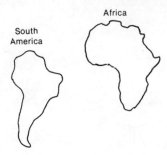

Millions of years ago, Wegener asserted, South America and Africa had been united.

They had since drifted apart to their present positions.

Wegener's theory of continental drift had been called "impossible" by a meeting of geologists in 1928, but by the early 1960s many geologists were asking: If islands could travel along the surface of the earth, why not continents? The spreading of the ocean floor on both sides of the Atlantic Rift, they continued, could offer a possible explanation of the separation of South America and Africa.

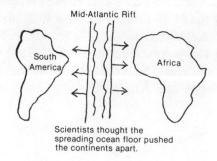

Scientists thought the spreading ocean floor pushed the continents apart.

The Journey to the Lost Continent

If the same kind of explanation could be applied to the separation of India, Australia, and Antarctica as well, then Gondwanaland could have existed even though there had never been any land bridges. But there were no known rifts—gigantic cracks in the earth—in either the Indian or Pacific oceans.

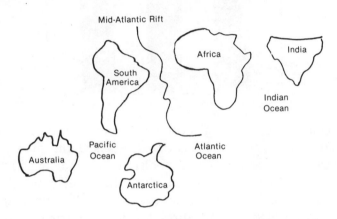

To provide evidence that Gondwanaland could have been broken up into five separate continents, rifts like the one in the Atlantic had to be found in both the Indian and the Pacific oceans. Echo sounding did discover a rift in the Indian Ocean. Was there also one in the Pacific? The Pacific is the world's largest ocean with an area of about 64 million square miles, and echo-sounding surveys would take years to find an answer. Geologists looked for a faster way. Two British geologists, Fred Vine and Drummond Matthews, found that way by solving a puzzle that had baffled geologists for more than a decade.

The Puzzle of the Deep-Sea Zebra Stripes

The earth is a giant magnet (a body having the property of attracting iron). The strength of the magnetism it produces can be measured by an instrument called a magnetometer. Beginning in 1952 geologists had measured the magnetism on the floor of the Indian Ocean by towing magnetometers behind ships. The measurements were surprising. Usually, the magnetic strength at one place on a magnet is very much like the magnetic strength at any other place. But what geologists found were alternating bands of weak and strong magnetism along the ocean floor. A map of these magnetic bands looked like the skin of a zebra.

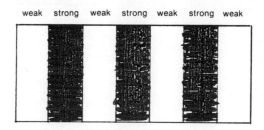

Until 1963 no geologist could explain this strange magnetic pattern. Then Vine and Matthews attacked the puzzle with science's two most powerful tools, reason and imagination. They started with the fact that when the magnetometer passes over rock that does not contain compass needles, nonmagnetic rock, the instrument measures the average strength of the earth's magnetic force pointing north. This measurement is known as the normal magnetometer reading.

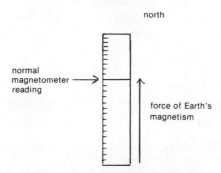

The strong bands of magnetism in the pattern of the deep-sea zebra stripes showed higher than normal magnetometer readings. Vine and Matthews interpreted that as meaning the magnetometer had passed over rock with compass needles pointing north. The magnetism of the rock had added to the earth's normal magnetism to produce a higher than normal reading.

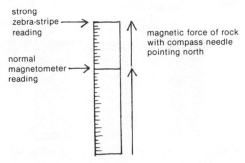

The weak bands of magnetism in the pattern of the deep-sea zebra stripes showed lower than normal magnetometer readings. To Vine and Matthews that meant the magnetism in the rocks the magnetometer passed over was subtracting from the earth's normal magnetism. The compass needles in these rocks, the British geologists concluded, must be pointing south.

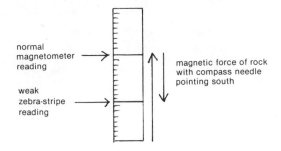

Each stripe of strong magnetism indicated the presence of compass needles pointing north,

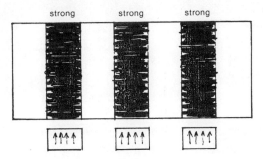

and each stripe of weak magnetism indicated the presence of compass needles pointing south.

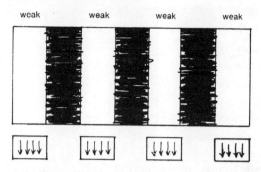

The pattern of zebra stripes on the floor of the Indian Ocean corresponded to a flip-flop pattern of compass needles.

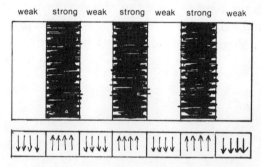

Moreover, the flip-flop pattern spread out on both sides of the Indian Ocean Rift.

J. Tuzo Wilson had interpreted a similar flip-flop pattern as evidence of ocean floor spreading. Independently of Wilson, Vine and Matthews concluded that the floor of the Indian Ocean was spreading on both sides of the Rift. The puzzle of the deep-sea zebra stripes had been solved.

From the solution of the puzzle it became clear to geologists that if the flip-flop pattern—the deep-sea zebra-stripe pattern—were found on an ocean floor anywhere, the presence of a rift would be indicated. Here was a fast way to find whether there was a rift in the Pacific: Look for the deep-sea zebra-stripe pattern on that ocean's floor. Research teams in laboratory ships with magnetometers atow found that pattern.

Geologists in search of the lost continent of Gondwanaland now had evidence of rifts in the Atlantic, the Indian, and the Pacific oceans. Since ocean floor spreading from both sides of those rifts could have broken up Gondwanaland and pushed the present-day southern continents apart, these geologists felt they were on their way toward proving that Gondwanaland had existed. They would feel certain they were on the right trail when they actually located the Pacific rift. But where were they to look for it in that vast expanse of water?

Two American geologists, Bruce Heezen and Maurice Ewing of the Lamont-Doherty Geological Observatory, found a clue that would supply the answer. That clue would also lead to a geographical discovery more amazing than Marie Tharp's.

The Clue of the Mirror-Image Pattern

Research teams led by the Scripps Institution of Oceanography in California had made detailed surveys of ocean floor magnetism all over the world, and had come back with accurate maps of deep-sea zebra patterns from every ocean. When Heezen and Ewing examined the maps, they found an exciting clue. The zebra-stripe patterns always formed on both sides of a band of extremely high magnetic strength. The pattern on one side was the mirror image of the pattern on the other, the same but opposite.

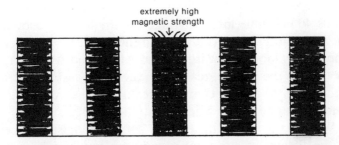

The clue was exciting because this was the same kind of pattern L. W. Morley had observed in the igneous rocks spreading out from the opposite sides of the Mid-Atlantic Rift.

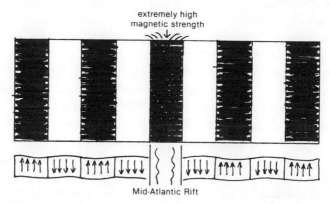

The band of extremely high magnetometer readings on the Atlantic floor coincided with the location of the Mid-Atlantic Rift. Wherever a band of extremely high magnetism appears on any ocean floor, Heezen and Ewing predicted, echo soundings will find a rift.

American and British echo-sounding expeditions searched for rifts along extremely high-intensity magnetic bands on ocean floors all over the world. When the findings of these expeditions were coordinated, an astonishing map was produced. Only a few years before, Marie Tharp had made history with her discovery of a rift-ridge system (an enormous cleft in the earth flanked by parallel mountain chains) running down the middle of the Atlantic floor. The new map revealed that the rift-ridge system Marie Tharp had discovered was only part of a continuous 40,000 mile long rift-ridge system winding around the globe. Beginning near Antarctica in the Atlantic, the main line of the system loops eastward around Africa and Australia, continues across the Pacific, reaches up the west coast of South and North America, swings through the Arctic, then connects with the northern end of the Mid-Atlantic Rift-ridge system.

With the Pacific Rift located, some geologists felt certain that the former existence of Gondwanaland had been established. The rifts in the Pacific, Atlantic, and Indian oceans, they reaffirmed, had been centers of ocean spreading that had pushed Gondwanaland apart. But other geologists were not so certain that this could have happened. They pointed out that

The Journey to the Lost Continent

Australia, the smallest of the continents that allegedly had been part of Gondwanaland, weighs more than 1,000 million million pounds, and an enormous force is necessary even to budge such a massive body. Could the ocean floor, these geologists asked, creeping along at the speed of about an inch a year, generate this kind of force?

By the late 1960s, when this question was asked, geologists had discovered that the ocean floor moves as separate rigid plates. Each plate (there are twelve major ones and a number of smaller ones) is associated with a specific section of the worldwide rift. The Pacific plate is associated with the Pacific Ocean Rift, the South American plate with the Mid-Atlantic Rift, the Australian plate with the Indian Ocean Rift, and so on. Geologists could determine the weight and speeds of these plates, and from that data they could calculate the force the plates exerted on the continents. No plate, geologists found, could produce a force capable of moving even such a small continent as Australia. That meant the five southern continents could not have been pushed apart by the ocean floor.

There was now no conclusive evidence that Gondwanaland had existed. The geologists journeying to the lost continent had turned into a blind alley. But in England a geologist was making a discovery that would lead to the solution of a 4-centuries-old mystery, and that solution would put those geologists back on the trail to Gondwanaland.

The Mystery of the Ring of Fire

The scene of the most intense earthquake and volcanic activity on this planet is a zone that encircles the Pacific Ocean.

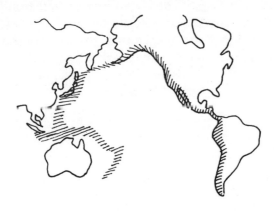

This zone is called the ring of fire, and it was charted by European explorers as early as the sixteenth century. But until 1967, no one could explain why most of the earth's most violent earthquakes and explosive volcanic eruptions happened there.

In that year Dan McKenzie, working at the geological laboratories of Cambridge University, built models of oceanic plates and continents and experimented with them to find out what would happen when plates collided with continents. Would the heavier continents stop the lighter plates and crumple or shatter them? McKenzie was surprised to find that the plates were not stopped. Instead, they were forced downward by the heavier continents. The descending plates cut deeply into McKenzie's models of the earth's interior.

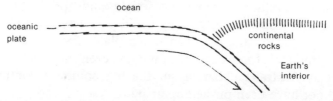

Did actual oceanic plates behave like those in McKenzie's models? Geologists had discovered that all along the ring of fire oceanic plates were colliding with continents. Research teams journeyed to numerous sites along the ring of fire to search for evidence that oceanic plates did cut into the earth. These teams were equipped with a special kind of echo-sounding equipment that sent sound waves through rocks. From the speed and direction of the echoes, geologists could draw a map of the interior of the rocks. These maps showed that the oceanic plates cut into the earth, producing gashes, called trenches, usually about 5 to 6 miles deep. Geologists could now explain in the following way the violent earthquake and volcanic activities of the ring of fire.

Earthquakes there are caused by the descending plates grinding against the continental rocks.

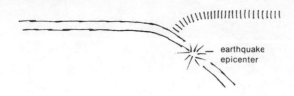

The severe shocks of these earthquakes, which occur below the surface of the ocean, are the cause of the feared tsunamis, commonly known as tidal waves, which race under the sea for about 2,000 miles at an average speed of about 450 miles an hour, then rise from the sea bottom and flood the land. (Elsewhere, earthquakes are caused by the collision, separation, or shearing of plates.)

Volcanoes along the ring of fire are caused by a chain of events. As the oceanic plate is forced down through the Moho into the hot inner crust, some of the rock begins to melt about 30 to 35 miles beneath the earth's surface, forming underground pockets of white-hot lava. The lava contains gases (mostly steam) under pressure, much as a bottle of a carbonated soft drink contains a gas (carbon dioxide) under pressure. When the bottle is shaken and then the cap removed, the gas erupts, carrying along with it a spray of liquid. Much the same thing happens to a pocket of white-hot lava. The pocket is shaken violently by

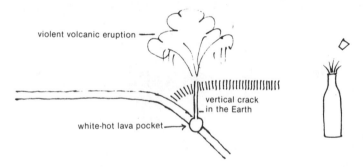

the motion of the descending plate. The cap on the pocket—the miles of continental rock exerting pressure from above—is removed when a vertical crack in the earth, caused by the collision of the plate with the continent, connects with the pocket. The pressure is suddenly released, and the gases explode, carrying the lava along with it. (A volcanic eruption along the rift system is gentle by comparison.)

With the discovery of the descending plates, the mystery of the ring of fire had been solved. But the solution of that mystery produced a new mystery. It would lead geologists to the ring of fire along the Pacific coast of South America, as they once again picked up the trail to Gondwanaland.

The Mystery of the Calm Coast

The Pacific plate in collision with the west coast of South America produced a zone of active trenches, with violent volcanoes and earthquakes.

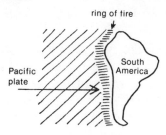

But the South American plate in apparent collision with the east coast of South America produced a zone without trenches, volcanoes, or earthquakes of any kind.

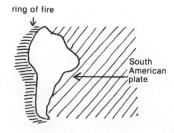

Why was the east coast of South America calm?

Dan McKenzie suggested an answer based on his study of plate models. The South American plate does not collide with South America; South America is embedded in the plate and is carried along by the plate.

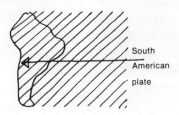

Here was a way out of the blind alley in which geologists in search of Gondwanaland found themselves when they discovered that the moving ocean floors could not have pushed the five southern continents apart. If McKenzie's suggestion proved correct, there was another way to account for the separation of at least one of those continents from Gondwanaland: South America had been carried away by the South American plate. To test McKenzie's suggestion, these geologists journeyed to the ring of fire to determine whether South America was actually moving west over the descending Pacific plate. Careful measurements of the position of the coastline over a period of time showed a continuous westward movement. McKenzie's suggestion had proved correct.

Were the other southern continents also carried along by plates? Earthquake-wave studies showed the existence of plates under all the continents of the earth, and all the plates were in motion. The plates rode on the solid-liquid rock of the earth's inner crust (which geologists now called the mantle). The force that propelled the plates, many geologists agreed, came from the convection currents that J. Tuzo Wilson had discovered in the earth's interior.

Geologists now knew that the five southern continents had drifted apart. But there was still no evidence that they had drifted apart from a common starting place, Gondwanaland. To find out, geologists would have to trace the paths of the moving plates back into time. That could only be done if the moving plates had left tracks—records on the ocean floor—of those paths. But what kind of track does a moving plate make? No one knew. Before geologists could search for the tracks, they had to discover the kind of track to search for. A simple clue led them to the discovery.

The Clue of the Misdirected Compass Needles

Geologists knew that in many samples of oceanic plate brought up by deep-sea drills, the compass needles pointed in directions other than north or south (the geographical positions of the north magnetic pole over the last several hundred million years).

But misdirected compass needles are an impossibility. Compass needles must point to the north magnetic pole. Since the compass needles in the oceanic plate could not have changed directions, geologists reasoned, then the plate must have changed directions. The many directions in which the compass needles pointed indicated that the plate had changed directions many times.

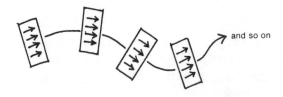

The changes in directions of the compass needles in an oceanic plate, they concluded, are the tracks of that plate.

The search for the tracks of the missing plates was conducted from the *Glomar Challenger,* one of the strangest ships ever to put to sea. Amidships sat a 142-foot-high oil drilling rig. Through a 20-foot-wide hole in the bottom of the ship, a miles-long string of steel drilling pipes were guided with great precision by sonar-commanded computers to drill sites on the ocean floor. There the drilling bits cut thousands of feet into bedrock. Long cores of the oceanic crust containing samples of the plates were pulled up to the ship.

Staffed by researchers from the Scripps Institution of Oceanography and four other American institutes, the *Glomar Challenger* collected the tracks of oceanic plates from all over the world, beginning in 1968. (The *Glomar Challenger*'s task, known as the Deep Sea Drilling Project, was to continue through 1975, and to be termed, according to the American oceanographer David C. Roberts,, "the most successful experiment ever done." It would enable geologists to reconstruct with great detail the precise sequence of events that led to the shape and positions of the present-day land masses.)

When the tracks the *Glomar Challenger* brought back in 1968 were dated by the potassium-40 clock, geologists were ready to trace in a general way the path of any plate back into time.

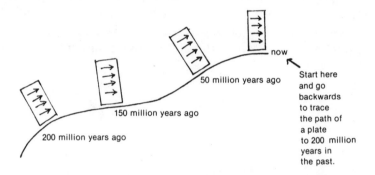

The Lost Continent

When the paths of the plates carrying the five southern continents were traced in this way, geologists found that all had started to move away from the same general location about 200 million years ago. Geologists also found that the moving plates had twisted some of these continents clockwise and others counterclockwise, so that the coastlines of these continents as they appear on a modern map look as if they had never fitted together. But by moving models of the five southern continents back along the paths of the moving plates, and reversing the twists, geologists showed that about 200 million years ago, all five had been joined smoothly into a single landmass—Gondwanaland.

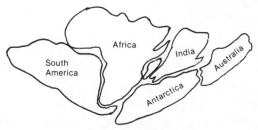

But, they were amazed to discover, Gondwanaland was not a continent. It was part of an even larger continent. Tracing the paths of the plates carrying the continents of North America and Eurasia (Europe and Asia), they had found that at that time these continents had been joined to each other and to Gondwanaland to form a single supercontinent. They called it Pangaea (Greek for all lands). This was the lost continent—the continent from which all the present-day continents were born.

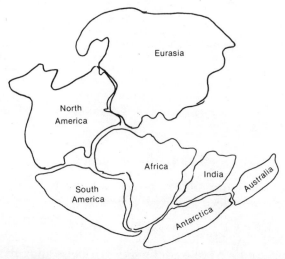

About 200 million years ago, according to a reconstruction made by the American geologists Robert S. Dietz and John C. Holden, some still unidentified cataclysm split open the earth's crust, producing the worldwide rift and severing Panagea into seven continents. The continents were then set adrift on the oceanic plates formed from the lava rising out of the rift. The moving plates sometimes collided with one another. When that happened, mountains were pushed up, as when the Indian plate rammed into the Eurasian plate and produced the Himalayas. (Mountain climbers reaching the peak of the highest mountain in the world, Mount Everest in the Himalayas, stand on what was once an ocean floor.) Twisting and turning, the ocean plates moved the continents (and all other land masses) to their present positions. And the continents are still adrift. (The study of the surface movement of the earth is called plate tectonics. *Tectonics* is a branch of geology concerned with the earth's structure.)

What will happen to the position of the continents in the future, assuming the plates continue in their present directions? North and South America will stay on a westerly course, but will break apart as Panama and Central America are pushed northward. Africa will close in on Europe, shrinking the Mediterranean to the size of a small lake. India for a while will burrow under Asia, pushing the Himalayas to even greater heights, but then it will turn eastward. Australia will voyage north, sideswiping Asia, which will slide east into a narrowing Pacific. And Antarctica will pull away from the South Pole.

The continents will change their shapes as the parts that lie on separate plates pull away from one another. The southern part of California, in which Los Angeles is located, for example, will tear away from the mainland and head northwest. Traveling at the rate of about 2 inches a year, it will sail past San Francisco 10 million years from now and reach Alaska 40 million years later.

With Pangaea as their starting point, geologists could neatly explain the changing geography of our globe from the remote past to the distant future. It would seem that the lost continent had been found. But many geologists doubted it. Pangaea, they held, could only have existed if continental drift—the movement of the continents on the surface of the earth over the last 200 million years—was true, and it had not been proven. The so-called proof, they said, was built on indirect evidence, such as magnetic patterns on the ocean floor, which could or could not have something to do with continental drift. Until direct evidence—facts that could be explained *only* by continental drift—was found, these geologists would continue to regard Pangaea as science fiction, not science fact. To find that direct evidence, Edwin H. Colbert journeyed to Antarctica.

The Find on Coalsack Bluff

Colbert, an authority on animal fossils, knew that in 1967 an intriguing fragment of bone had been found on a mountain near the South Pole. The bone was identified as coming from the jaw of a labyrinthodont, an animal that had lived about 200 million years ago. Labyrinthodont remains had previously been found in Africa. Colbert knew that the animal could not possibly have swum to Antarctica because it was unable to live in saltwater. Colbert thought that if remains of other non-ocean-going animals typical of Africa at that time could be found in Antarctica, it would be direct evidence that these continents had once been joined together and then drifted apart.

A fossil-hunting expedition headed by Colbert set up camp at Coalsack Bluff on a mountain close to where the labyrinthodont find had been made. In November 1969 Colbert wrote to his associate Bobb Schaeffer of the American Museum of Natural History in New York:

> Dear Bobb,
> We did it!
> On our first day of field work we found a cliff full of Triassic reptile bones . . .

These bones of reptiles that had lived about 200 million years ago, as well as those later found by the Colbert expedition, were virtually identical with similar specimens found in Africa. Among the bones unearthed by the Colbert expedition were those of the lystosaurus, a reptile that resembled a rhinoceros, and of the thrinaxodon, a weasellike reptile. Neither of these reptiles, which were freshwater dwellers, could have swum across the 2,400 miles of saltwater that now separate South Africa from Antarctica. "This really pins down continental drift," Colbert added in his letter to Schaeffer.

The find at Coalsack Bluff, by supplying direct evidence of continental drift, put beyond doubt the former existence of

Pangaea. At Coalsack Bluff science's journey to the lost continent came to an end.

But at journey's end, a host of mysteries remained, including: Do convection currents generate sufficient energy to drive the plates? (Some geologists think not.) And if not, is there a powerful unknown force operating inside the earth? Why are parts of Africa breaking off and moving away faster than the mainland, even though all are on the same plate? Why do some volcanoes in the Hawaiian Islands continue to spew out lava even after their pipelines to the molten rock below the Moho have been severed by moving plates? How can an earthquake be predicted with accuracy? What clues are there in the plates that would lead to the location of oil, gas, and mineral resources?

How will the answers be found? Perhaps one by one, as the technologies of deep-sea drilling and echo sounding improve. Perhaps in one great new quest that will solve all the mysteries with a single basic discovery. But either way, geologists will move forward in their continuing journey to uncover the secrets of the earth.

Part II—The Universe

3

The Journey to the Strange World Inside the Atom

The world learned about the first journey to the interior of the atom from a mysterious event that occurred in 1936. On September 29 of that year, at 3:31 P.M., an unidentified flying object, so huge that it blotted out the sun, struck Lake Erie a few miles east of Cleveland, Ohio. The impact produced a sound that was heard for miles and caused the waters of the lake to rise with a rush. At least one third of Cleveland was flooded, and devastating damage was reported in lake cities as far east as Erie, Pennsylvania, and as far west as Toledo, Ohio.

According to thousands of eyewitnesses, the unidentified flying object had the shape of a man and appeared to be shrinking at an extremely fast rate. Within hours of its appearance no trace of it could be found. Dr. Hilton U. Cogsworthy of the Alleghany Biological Society, in a statement issued to the press, asserted that the object eyewitnesses had described could not possibly have existed. He concluded that the sighting was "the result of some kind of mass hypnosis on a gigantic scale."

On the night of the sighting a stranger walked through an open doorway of a house in the suburbs of Cleveland. He was 11 feet tall and shrinking rapidly. He communicated by means of brain waves with the owner of the house, Stanton Cobb Lentz, the renowned author of *The Answer to the Ages*. He told Lentz the story of the incredible events that had led up to the splashdown. Following is a digest of the stranger's story as it appeared under Lentz's by-line in the *Cleveland Daily Clarion*.

I was chosen (the stranger said) to be injected with a new drug which would cause me to shrink continuously without ever stopping. No matter how small I would become, I would always become smaller. Shortly after being injected with the drug, I became smaller than a pencil and I was placed on a sheet of the metal Rhyllium-X. I continued to shrink and slipped into the smallest particle of Rhyllium-X, the atom. I was surprised to find that the atom resembled our solar system. A central sun was surrounded by many orbiting planets.

As I burst into this solar system, I was gigantic compared to the planets. But I shrank, and I was soon able to alight on one of them. The inhabitants of the planet were super-intelligent sphere-shaped creatures. They captured me and handed me over to their scientists who observed me as if I were a strange kind of animal. I went on shrinking, and when I became too small for them to see me, I was placed on a slide under a microscope. They watched me until I became so small that I vanished into an atom of the microscope slide.

This atom was a solar system much like the last. When I landed on one of its planets, I was still many hundred times the size of the primitive jungle people who lived there. I rescued a tribe of them from a huge dragon-like creature by swatting it as if it were an insect on my own planet. The tribe worshipped me like a God. But in a short time, I was smaller than the people I had rescued and I was forced to flee from monstrous beasts like the one I had killed so easily. Finally, I became so small I fell into an atom of sand.

Once again, I found myself floating in a solar system. The planet I touched down on this time was dominated by a race of evil machines. The former masters of the planet, a race of gentle bird-like people, had fled to a nearby moon. The evil machines were about to launch a war against the bird-like people and all other living things in the solar system. I escaped from these terrible machines by shrinking into an atom of one of the few remaining blades of grass on the planet.

After that, I shrank through twenty-nine solar systems, each an atom, and each atom smaller than the previous one. My thirtieth atom was also a solar system. I was attracted to a tiny blue planet. I arrived on that planet on September 29, 1936, at 3:31 P.M., splashing down in Lake Erie a few miles east of Cleveland, Ohio. I had been the unidentified flying and shrinking object that had baffled your scientists here on Earth.

Lentz reported that when the stranger finished his story, he was about 2 inches in height and was "steadily and surely diminishing." Lentz added, "A memory that will live with me always is the sight of [the stranger] as last seen by me—as last seen on this earth . . . waving two arms upward as in farewell . . . as [he] disappeared into infinite smallness." THE END.

The incredible tale you have just read is a condensed version of *He Who Shrank,* a short novel written by the American author Henry Hasse in 1935. The basic idea of the novel is based on one answer to a problem that had intrigued Greek philosophers more than 2,000 years before. The problem was: What would happen if we took a piece of matter and kept cutting it into smaller and smaller pieces.

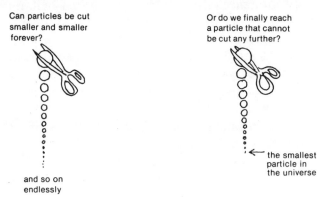

Some ancient Greek philosophers answered that no matter how tiny a piece became, it could always be cut to a smaller piece. They believed that any small piece—that is to say, any particle—of matter of any kind always contained a smaller particle. Hasse used that idea to have his stranger shrink into smaller and smaller particles—atoms—without ever reaching a particle that did not contain a smaller one.

Other ancient philosophers answered the problem by saying that no matter how sharp the cutting tool, it would not be possible to cut up a particle smaller than a certain size. That smallest particle in the universe would contain no other particle. The Greeks called it a fundamental particle, meaning that it was a basic building block of matter. They gave it a special name, atom, which in Greek means "that which cannot be cut any further."

It wasn't until 1805 that scientific evidence was found of the existence of basic building blocks of matter. They were called atoms. But the name was wrong. About a hundred years later, physicists found that the atoms science had discovered could be cut—actually, broken down into the smaller parts they contained. The search then started for particles that could not be cut, the true fundamental particles.

The search took physicists into the interior of the atom. (Physicists are scientists concerned essentially with what things are made of, and what holds them together and pulls them apart.) They journeyed there by means of brilliant experiments, bold feats of reason and imagination, cleverly designed apparatus, and gigantic machines unlike anything ever built on earth.

This is the story of that journey. It is a story filled with wonders stranger than any encountered by Hasse's shrinking man. Inside the atom physicists found a world of bizarre particles to which they gave queer names such as leptons, muons, and quarks. And they found forces that physicists of only a couple of decades ago never dreamed could exist. But did they find the fundamental particles? Like all stories of scientific discovery, this is a mystery story. And like the best of mystery stories, this one has a surprise ending.

The story begins with man's first journey to the atom. John Dalton, an English schoolteacher, found his way there by solving a puzzle that had baffled scientists for years.

The Puzzle of the Perfect Proportions

When the nineteenth century began, chemists were trying to discover what things were made of. (Chemists are scientists specializing in making new substances, as well as breaking down substances into their parts.) For example, what was sugar made of? Let's break it down and see, they decided. They devised a simple experiment to start them off. They heated a spoonful of sugar over a flame. The sugar turned into a sticky black substance that could not be broken down any further. They called it carbon.

They repeated the experiment. This time they held a cool knife blade over the heating sugar. Moisture formed on the blade. In addition to carbon, sugar contained water.

But what was water made of? They tried another experiment to find out. They connected wires to an electric battery and placed the ends of the wires into water. The electric force caused bubbles of oxygen gas to come out of the water at one wire and bubbles of hydrogen gas to come out at the other. Since these gases could not be broken down into anything else, they concluded that water was made of hydrogen and oxygen.

So sugar was made up of carbon, hydrogen, and oxygen. These components of sugar were called elements—substances that nineteenth-century scientists believed could not be broken down into other substances. In the first years of the 1800s, twenty elements were known. (Since then about another eighty have been identified.)

Elements combined with each other to produce substances called compounds. The elements hydrogen and oxygen, for example, combined to form the compound water. Scientists believed that elements, and elements in the form of compounds, made up all matter. But what were elements made of?

In 1805 John Dalton set out to find the answer. His clue was a puzzle having to do with how much of each element was needed to make a compound. Here is what the puzzle was all about as it applied to the compound water.

Chemists of Dalton's time could produce water by mixing oxygen and hydrogen gases in a strong tank, then exploding the mixture by means of a spark. But water would only form when 1 ounce of hydrogen was mixed with 8 ounces of oxygen,

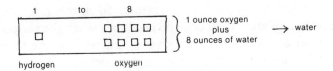

or when 2 ounces of hydrogen were mixed with 16 ounces of oxygen,

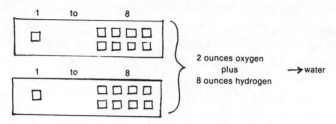

or when 3 ounces of hydrogen were mixed with 24 ounces of oxygen,

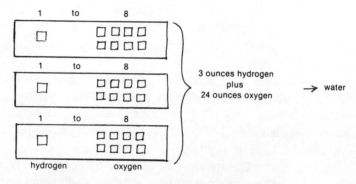

and so forth.

In every case the basic relationship by weight between hydrogen and oxygen was always one to eight.

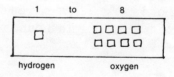

The relationship of any quantity to another is called a proportion. Water could only be produced when the proportion by weight of hydrogen to oxygen was a perfect one to eight. Why did the proportion have to be perfect? That was the puzzle.

The Journey to the Strange World Inside the Atom

Dalton solved it with a superb display of reasoning. He imagined two balls of equal volume, one made up of hydrogen and the other of oxygen. He knew that a volume of oxygen was sixteen times heavier than an equal volume of hydrogen, so the ball of oxygen would be sixteen times heavier than the ball of hydrogen.

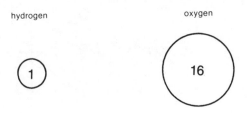

If, Dalton went on, it were possible to shrink these balls uniformly, then no matter how small the balls became,

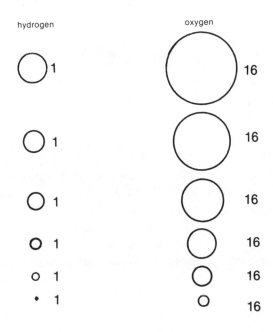

the ball of oxygen would always weigh sixteen times more than the ball of hydrogen.

Let's suppose, Dalton said, we shrink these balls down so small that they become invisible. When we mix one of these balls of hydrogen with one ball of oxygen, we can't produce water

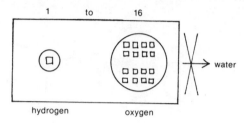

because the proportion by weight of hydrogen to oxygen is one to sixteen, not one to eight. But if we mix two balls of hydrogen with one ball of water, the proportion is now

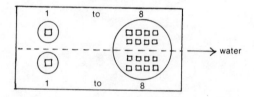

a perfect one to eight, and water is produced.

It seemed clear to Dalton that hydrogen was composed of tiny balls of a specific weight, and oxygen of tiny balls of a different specific weight. This idea solved the puzzle of the perfect proportions as it applied to the production of water.

Dalton then extended his idea to the production of other compounds. He knew that all elements combine in perfect proportions to form compounds. Dalton concluded that all elements were composed of tiny balls that differed in weight from element to element. These tiny balls in various combinations accounted for the perfect proportions by weight needed to produce compounds.

Dalton said that the tiny balls of an element could not be cut any further, because if they could, then their weights would

be less than those required to provide the necessary perfect proportions. He called the tiny balls he discovered atoms. Ever since, scientists have considered atoms to be the smallest units that show the properties of an element.

By a brilliant feat of reasoning Dalton had journeyed to the atom. Atoms are so small that if people were the size of atoms, the entire population of the United States, about 225 million people, would fit on the head of a pin. Dalton couldn't see atoms, but he pictured them as little balls without parts. Because scientists had no way of getting into an atom to find out if it had parts, they went along with Dalton's picture. Then, about 90 years later, during a rainy week in Paris, a French physicist made a chance discovery. It would trigger a chain reaction of discoveries that would end with science's first journey to the inside of the atom—and Dalton's picture of the atom would be shattered.

The Discovery of a Strange Activity

It was 1896. The year before, William Konrad Roentgen, a German physicist, had discovered X rays. They were strange new rays that could go right through matter—and the discovery caused a great sensation. Scientists began to look for more sources of X rays. One of these scientists was the Frenchman Antoine Henri Becquerel. He believed that fluorescent substances might just be such sources.

Fluorescent substances give off an eerie glow after they have been exposed to sunlight. Becquerel thought up an experiment to determine whether this fluorescent glow contained X rays. He wrapped a photographic plate with heavy black paper so that no light could reach the plate. Then he exposed a sample of fluorescent metal to sunlight, and placed the softly glowing metal on the paper. If the plate should fog, Becquerel reasoned, then the glowing metal would be giving off rays that could go right through the black paper—in other words, X rays. Becquerel unwrapped the plate and examined it. It had fogged.

Becquerel thought he had discovered a new source of X rays. But to be sure, he had to repeat his experiment. Then it rained. Without sunlight the metal would not fluoresce. And without fluorescence, there could be no repeat experiment. It rained for three days. Becquerel became impatient. Even though the metal was not fluorescing, he placed it on a wrapped photographic plate. Because the metal didn't glow, he expected to see the plate unaffected. But when he unwrapped the plate, he saw that it had fogged. For the rest of the week as it continued to rain, Becquerel ran his experiment with the non-fluorescing metal over and over again. Every experiment gave the same result as the first. There was no doubt in Becquerel's mind: The metal was sending out invisible rays.

The rays that fogged Becquerel's photographic plates had nothing to do with fluorescence. The metal did not have to be stimulated into activity by sunlight. The metal itself was active. Physicists called this strange activity radioactivity (from the Latin word *radiare,* meaning "to send out rays"). The metal that fogged Becquerel's plates was the first radioactive substance known to science. It was a compound of the element uranium.

Uranium forms compounds with various elements, and each uranium compound Becquerel tested was radioactive. Rays from them behaved like X rays in that they could pierce solid matter. But they differed from X rays in other respects. (For example, physicists could take pictures of the insides of solid objects with X rays, but not with the rays from uranium compounds.) Becquerel had discovered new rays. Scientists wanted to know more about them. But the rays from the uranium compounds were too feeble to be studied easily. A stronger source of radioactivity was needed.

The Search for a Powerful Source of the New Rays

By simple reasoning, physicists had identified one such source. The other elements in uranium compounds existed in other compounds that were not radioactive. That meant that uranium was the source of radioactivity. If pure uranium could be obtained, physicists would have a powerful source of radioactivity with which to conduct their experiments. But pure uranium does not exist in nature. How could it be obtained?

The problem fascinated Marie Curie, a young Polish-born chemist who was busy cooking for her husband, looking after her infant daughter, and thinking about what kind of research to do. With the collaboration of her husband, Pierre, a French physicist, she set out to isolate a sample of pure uranium.

Uranium is found in the crust of the earth in compounds locked up in an ore called pitchblende. An ore is a mixture of compounds of a certain metal and other substances. These other substances in pitchblende are nonradioactive. Marie Curie's first job was to get rid of the nonradioactive substances. To do it she had to use slow, tedious methods devised by chemists for separating substances from one another. (The principal method was fractional crystallization, a process by which crystalline compounds of a substance are thrown out of solution—that is, precipitated—in small fractions of the total amount of the substance dissolved. To obtain all of the dissolved material, the crystallization frequently has to be repeated several hundred times. Each fractional crystallization yielded Marie Curie extremely small amounts of radioactive material.) She started with a ton of uranium ore. It took her 2 years of painstaking work to whittle it down to a radioactive sample weighing less than an ounce. From that sample, she separated pure uranium.

The task she had set for herself was completed. But her work was not. Something extraordinary had occurred. After she had separated the pure uranium from her sample, an extremely small amount of material had been left over. She thought it would be nonradioactive. She was astonished to find that it was even more radioactive than uranium. From the leftover material she separated a new element, the second radioactive element known to science. It was four hundred times more radioactive than uranium. She named it polonium in honor of her native Poland.

Now no more than a speck—about .01 ounce—of unidentified material remained from the ton of pitchblende with which she had begun her project. That speck was like no other substance ever before seen in a laboratory. It generated its own heat and glowed faintly in the dark. It was far more radioactive than polonium. Madam Curie concluded that still another radioactive element produced that radioactivity, and she named it radium. She had discovered an element about a million times more radioactive than uranium.

The year was 1898. Now physicists had a powerful source with which to study the mysterious new rays Becquerel had discovered. No one at that time thought that these studies would lead to a technique for smashing into the atom. What concerned physicists then was: Just what were Becquerel's rays made of?

A Clue from a Radium Gun

Ernest Rutherford, an English physicist who was to become the great pioneering explorer of the atom, knew that if a stream of particles passed near a magnet, the stream would be bent. If the new rays were bent by a magnet, then Rutherford would know the rays were made up of particles. He set out to find what effect a magnet would have on a stream of rays from a speck of radium.

Since the rays from the speck flew off in all directions, like light from a miniature sun, Rutherford's first task was to guide them into a stream. He invented a device to do it—a radium gun.

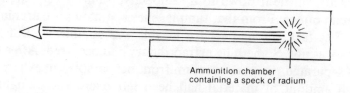

Ammunition chamber containing a speck of radium

The gun was made of lead because rays from radioactive substances cannot pass through this heavy metal. The rays were channeled into a stream by the gun barrel. When a lead stopper was removed from the opening of the barrel, the rays streamed out in a straight line.

Rutherford loaded his radium gun and stopped it. He positioned three photographic plates opposite the gun and placed a magnet between the gun and the plates.

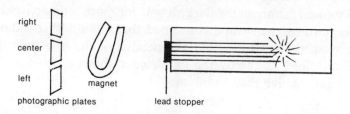

The Journey to the Strange World Inside the Atom 67

One plate was in a direct line with the stream of rays. Another was to the right of that line, and the third was to the left. If the rays were made of particles, the right or the left plate would be fogged.

Rutherford removed the lead stopper and found to his astonishment that

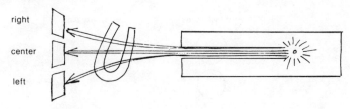

all three plates were fogged. That meant the speck of radium was shooting out three different kinds of rays. The magnet had bent one kind to the right and one to the left; it had not bent the third. Rutherford named the rays alpha, beta, and gamma, respectively (after the first three letters of the Greek alphabet).

He concluded that the rays that had not been bent, the gamma rays, were not made of particles. But

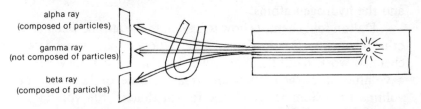

the rays that had been bent, the alpha and beta rays, were made of particles.

Now Rutherford wanted to know: What kind of particles? He teamed up with an English chemist Frederick Soddy to find the answer. In a short while, they came across an unusual clue.

The Clue of the Atoms That Lost Weight

Physicists didn't know the actual weights of atoms, but they did know the weight of an atom of an element as it compared with the weight of an atom of oxygen. About 100 years before, Dalton had worked out a way to obtain these comparative weights. He started by weighing a volume of oxygen and then an equal volume of hydrogen. The volume of oxygen was sixteen times heavier than the volume of hydrogen. Dalton knew that when these volumes were shrunk uniformly to the volume of an atom, the relative weights would remain unchanged.

any volume of hydrogen | same volume of oxygen

○ 1   ○ 16
○ 1   ○ 16
○ 1   ○ 16
○ 1   ○ 16
○ 1   ○ 16
• 1   • 16

Dalton assigned the number 16 to the atom of oxygen and the number 1 to the atom of hydrogen. These numbers, which he called atomic weights, stood for the weights of the oxygen and the hydrogen atoms.

Dalton told chemists how to calculate the atomic weights of other elements. For example, here is how, many years later, chemists worked out the atomic weight of helium. They weighed a volume of oxygen, then an equal volume of helium. The volume of helium weighed one-fourth that of the volume of oxygen. That meant helium had one-fourth the atomic weight of oxygen.

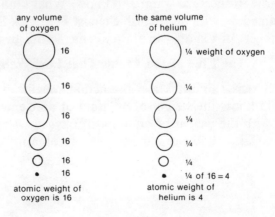

The Journey to the Strange World Inside the Atom

Since the atomic weight of the oxygen atom had been set by Dalton at 16, the atomic weight of helium turned out to be 4 (one-fourth of 16). (Actually, the atomic weight of helium is 4.0003, but to keep things simple, this atomic weight and all others are given in round figures.)

Rutherford and Soddy knew the atomic weight of radium was 226. But when radium was sealed for some time in a lead container, they found traces of a radioactive gas with an atomic weight of 222. No known element had that atomic weight. The two English scientists had discovered another radioactive element, which they called radon. It was obvious to them that some radium had changed into radon, and in the process some radium atoms had lost four units of atomic weight (226 minus 222). Where had the lost weight of the radium atoms gone? It was possible, Rutherford and Soddy thought, that particles weighing four atomic-weight units had been ejected from the radium atoms in the form of particle rays.

Here was a clue to the makeup of the alpha and beta rays, and Rutherford and Soddy followed it up. They trapped alpha rays in a lead container. When they examined the contents of the container, they found a very small amount of nonradioactive gas. It had an atomic weight of 4. Since helium had an atomic weight of 4, Rutherford and Soddy concluded that alpha rays were made of helium atoms.

It was clear to Rutherford and Soddy that an atom of radium had split into atoms of two other elements, helium and radon.

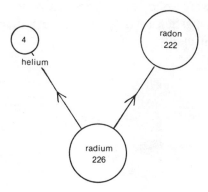

This discovery, which was made in 1903, rocked the scientific world. For about 100 years physicists had believed Dalton's idea that an atom could not be cut any further. Now Dalton had been proven wrong. Had Dalton also been wrong when he pictured an atom as a little solid ball with no parts? To find the answer, physicists had to find a way to look for invisible parts in an invisible atom. It looked like an impossible task. Rutherford found a way 8 years later. It would lead him to where man had never gone before:

To the Inside of the Atom

Rutherford journeyed to the inside of the atom by means of a superbly simple experiment. He set up a sheet of gold foil in front of his radium gun. Behind the sheet he placed a screen coated with a chemical that emitted tiny sparks when alpha particles struck it.

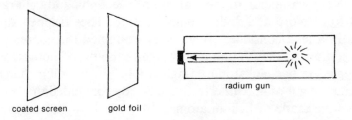

coated screen   gold foil   radium gun

The gold atoms (atomic weight 197) were far heavier than the alpha particles (atoms of helium, atomic weight 4). If the gold atoms were solid, as Dalton had pictured them, then the lighter alpha particles would bounce off them and the screen would remain blank.

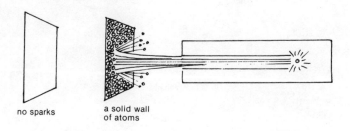

no sparks   a solid wall of atoms

But if the balls were not solid, then the alpha particles would pass through them and show up as sparks on the screen.

Rutherford removed the lead stopper from his radium gun. The alpha particles sped against the gold atoms. Instantly, in the direct path of the alpha particles,

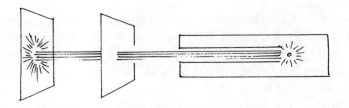

the coated screen was ablaze. The alpha particles were shooting right through the gold atoms. Dalton had been wrong. Atoms were not solid.

Was there anything inside the atoms? Rutherford thought his experiment would provide an answer. He knew that when a stream of water collides with a rock, it is diverted. Rutherford expected a stream of particles to behave in the same way if it were to strike a part of an atom.

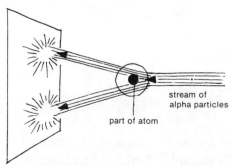

The diverted particles would show up as sparks on the screen in positions away from the direct path of the stream. But Rutherford saw no sparks in those positions. It seemed as if the alpha particles had gone through empty space.

Rutherford continued to watch the screen. He saw a spark near the top.

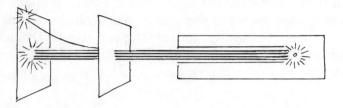

Then he saw another spark near the bottom. Other single sparks appeared to the right and left of center. Single sparks were caused by single alpha particles. Rutherford now knew that there was something inside the atom, and that that something was changing the path of a few alpha particles in the stream. But why was it changing the path of only a few particles and not the whole stream? In a sudden flash of insight Rutherford came up with an answer.

An atom, he explained, is a ball made up mostly of empty space surrounded by a shell. Most of the alpha particles in the stream passed through the shell and the empty space in straight lines,

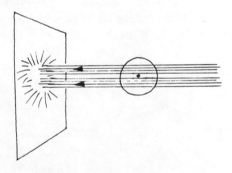

accounting for the burst of sparks at the center of the screen. But a few particles collided with a small object inside the atom

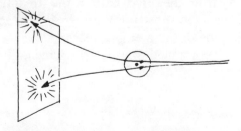

and bounced off that object in different directions. That accounted for the scattering of single sparks away from the center.

The chain of discoveries that had begun on a rainy week in Paris 15 years before with Becquerel's discovery of radioactivity had reached its climax. Rutherford had journeyed to the inside of the atom.

Rutherford called the object at the center of the atom the nucleus. He had made one of the most important discoveries in the history of mankind; in 1945 a massive effort of physicists, chemists, engineers, and civilian and military administrators would unlock the tremendous power of the nucleus, and the age of nuclear energy would begin. But in 1911, when he discovered the atomic nucleus, Rutherford was not concerned with the problem of energy. He was preoccupied with finding out more about the unexplored territory of the atom. His attention turned to the shell. Alpha particles seemed to go right through it as if it weren't there. What strange substance was the shell made of?

Rutherford found the answer in the path-breaking experiments of another English physicist, J. J. Thomson.

The Bottled Lightning Experiments

Thomson was interested in electricity. Electricity is a force that can push things apart or pull things together. A thing possessing an electric force is said to have a charge. There are two kinds of electrical charges: one is called positive (+) and the other negative (−). When things have the same kind of charge, the electric force pushes them apart.

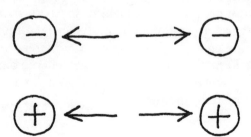

When things have different kinds of charge, the electric force pulls them together.

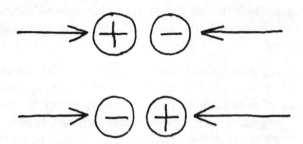

Thomson knew that if an object is heavily charged with the electric force, a spark will fly across the air to something close by. This is what happens when lightning flashes between clouds, or between clouds and earth, during a storm. Thomson decided to bottle a lightning flash. His bottle was actually a specially made glass tube. This is a simplified diagram of a cross-section of that tube.

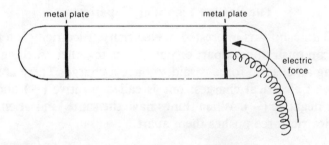

Thomson also knew that he could drive the electric force through a copper wire by attaching the wire to a battery (a chemical device for producing the electric force). Thomson connected one end of a copper wire to a battery and the other end to a metal plate in the tube. A heavy charge built up on the plate, and lightning flashed between the charged plate and the other metal plate in the tube.

The Journey to the Strange World Inside the Atom

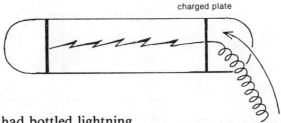

Thomson had bottled lightning.

It was obvious to Thomson that the lightning was caused by the action of the electric force on the air in the tube. If the air were removed, he reasoned, he would be able to bottle not lightning but the electric force itself as it flew between the two metal plates. The year was 1897, and by that time scientists had invented pumps that could extract most of the air from a container. So to remove the air from his tube would be no problem.

But what *would* be a problem—and a difficult one—was this: The air made the electric force visible in the form of lightning. Without the air in the tube, the electric force couldn't be seen. How would he know that the electric force actually flew between the two metal plates?

Perhaps, Thomson thought, the electric force might leave marks on the wall of the tube. He made a slit in the uncharged plate so that the electric force could pass through the plate and strike the wall of the tube.

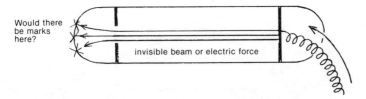

Then he pumped most of the air out of the tube and repeated the experiment that had produced the lightning flash. This time there was no flash, but a spot on the wall of the tube exactly opposite the opening in the metal plate shone with a blue-green glow.

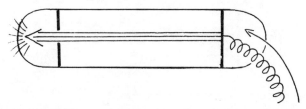

Thomson had trapped the pure electric force. Moreover, the slit had channeled the force into an invisible ray that followed a straight path. Was this ray composed of particles? If it were, then it would be deflected by a magnet. Thomson placed a magnet in the path of the ray. The magnet bent the ray to one side.

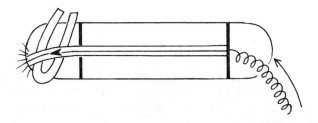

The electric force was composed of particles. Were the particles all of the same weight? If they were of different weights, the magnet would have bent the paths of the lighter particles more than the paths of the heavier particles.

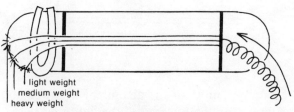

This is what would happen if the particles had different weights.

But that hadn't happened. The paths of all the particles had been bent equally. All had the same weight.

What was the charge of these particles, positive or negative? Thomson sandwiched the ray of electric force between two metal plates, one positively charged, the other negatively charged.

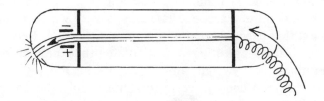

The particles were bent toward the positively charged plate. Since opposite charges attract, Thomson knew that the particles were negatively charged.

Six years before, the Irish physicist G. J. Stoney had predicted the existence of particles of electric force and had called them electrons. Thomson gave that name to the negatively charged particles he had discovered. The electric force, Thomson concluded, is a stream of electrons.

Fourteen years after the discovery of the electron, Rutherford was searching for the strange substance that seemed to make up the shell of the atom. He knew that atoms of radioactive elements shot out three rays. Since these rays came from the atom, Rutherford made the rational assumption that they were parts of the atom. Were any of these rays the strange substance of the shell? He ruled out gamma rays because he thought they contained no particles. (Physicists had found that gamma rays were much like X rays.) He also ruled out alpha rays because he thought they were made up of atoms (of helium), and the shell was not an atom but only a part of an atom. That left beta rays, which he knew were particles because their path had been bent by a magnet. Since he knew of no other parts of the atom except the nucleus, he concluded that beta ray particles made up the shell. But what were those particles?

From the published reports of Thomson's bottled lightning experiments Rutherford learned that the path of electrons was bent by a magnet. By comparing Thomson's data with his own, Rutherford found that a magnet bent the path of beta particles in exactly the same way as it bent the path of electrons. That meant beta particles *were* electrons. Rutherford had found the strange substance that made up the shell.

He explained how alpha particles shot through a shell of electrons as if it weren't there. Alpha particles were about eight thousand times heavier than electrons and traveled at a speed of about 10,000 miles a second. The energy with which a particle strikes another becomes greater the more the particle weighs and the faster it travels. Rutherford calculated that the high-speed, comparatively heavyweight alpha particles hit the atomic shell with tremendous energy, easily brushing aside the comparatively lightweight electrons. (Physicists had been able to compute the speed of alpha particles and the comparative weights of alpha particles and electrons by measuring how much the paths of these particles were bent by a magnet. Certain amounts of bending corresponded to certain speeds. The paths of light particles were bent more than the paths of heavy particles.)

Physicists now knew that the atom was made up of a nucleus and a shell of electrons. But what was the nucleus made of?

The First Exploration of the Nucleus

Physicists began their exploration of the nucleus by studying the hydrogen atom. With an atomic weight of 1, it was regarded as the simplest of all atoms. Several physicists working independently conducted an experiment such as the following one.

Virtually all the air was pumped out of a tube such as the one used by Thomson in his bottled lightning experiments. A small amount of hydrogen gas was then introduced into the tube. One of the metal plates was connected to a battery, and a stream of electrons bombarded the hydrogen atoms.

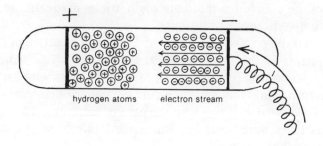
hydrogen atoms   electron stream

Physicists had found that the electrons in this stream traveled at many times the speed of alpha particles. At this high speed they would strike electrons in the atomic shells with sufficient force to knock them out of the hydrogen atoms,

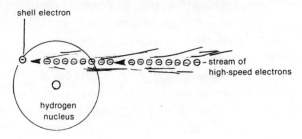

leaving the nuclei behind.

If the nuclei were electrically charged, they would be attracted to one of the oppositely charged plates in the tube. Examination of the negative plate showed it had attracted a small amount of matter. Nothing else in the tube could be attracted to the negative plate—electrons were negatively charged, and hydrogen atoms had no charge—so that small amount of matter had to be made up of hydrogen nuclei. And since objects with opposite charges attract each other, the nuclei had to be positively charged.

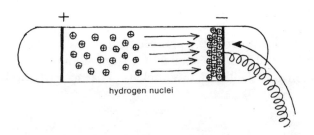

In separate investigations the hydrogen nucleus was found to have an atomic weight of 1, the same as that of the hydrogen atom. (The weight of an electron is negligible compared with the weight of a nucleus.) Because 1 is the first number in a scale of ascending atomic weights of the elements, the hydrogen nucleus was called the proton, which in Greek means "the first one."

The first exploration of the atomic nucleus had been completed successfully. Physicists went on to explore the rest of the hydrogen atom.

How many electrons were in it? To work out the answer, physicists applied their knowledge of objects that had no electrical charge, such as hydrogen atoms. They were called neutral objects. They could be made up of equal amounts of positive and negative charges, since opposite charges cancel each other. For example, one positive charge (+1) plus one negative charge (−1) equals no charge (0).

Physicists had found that the charge on the proton was equal (but opposite) to the charge on the electron. The hydrogen nucleus was a single proton. To produce a neutral hydrogen atom (0), the charge of one proton (+1) had to be canceled by the charge of one electron (−1), so there was only one electron in the hydrogen atom.

Physicists now knew that the hydrogen atom was composed of a proton and an electron. Where were these particles in relation to each other? Since they were oppositely charged, they should be attracted to each other. A map of the hydrogen atom should look like this:

But in Rutherford's pioneering journey to the inside of the atom, he had found a shell and a large amount of empty space between the shell and the nucleus. His map of the hydrogen atom looked like this:

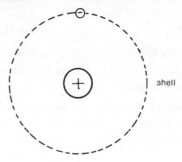

Which map was the correct one?

The Clue of the Upside-Down Pail of Water

If Rutherford's map were correct, why didn't the negatively charged electron and the positively charged proton attract each other? And how could one electron make up a shell? In 1913 a Danish physicist Niels Bohr found a clue to the answers in a children's game. The game was played with a pail of water attached to a string. The object of the game was to whirl the pail overhead upside-down without a drop of water spilling from it.

To play the game, children spun the pail overhead at high speed. The swift spinning motion generated a force that pushed the water against the bottom of the pail. This force balanced the force of the earth's gravity, which acted to pull the water out of the pail. So no water fell when the pail was upside-down.

Bohr concluded that the same kind of swift spinning motion that kept the water from falling to earth kept the electron from falling to the nucleus. His calculations showed that to generate the spinning force necessary to balance the force of electrical attraction, the electron whirled around the nucleus 7 billion billion times per second. At this ultrahigh speed, the electron wove a shell around the nucleus, much as the spinning blades of an airplane propeller formed an apparently solid disk.

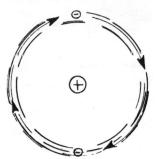

On the basis of Bohr's explanation, physicists accepted Rutherford's map of the hydrogen atom. (They soon discovered, though, that the explanation of the motions of a shell electron was more complicated than Bohr thought, and a new branch of physics called quantum electrodynamics was developed to describe these motions with great accuracy.)

The proton and the electron were regarded as fundamental particles. Physicists assumed that all atoms were made up of just those two particles. To test that assumption, researchers journeyed into the helium atom in search of protons and electrons—and came back with a new mystery.

The Mystery of the Missing Charge

The search began with an exploration of the helium nucleus. How many protons did it have? The researchers couldn't get into the nucleus to count them, but they thought the atomic weight of the nucleus would reveal the answer. An atomic weight of 1 meant one proton (as in the hydrogen atom); an atomic weight of 2, two protons, and so on. Since the atomic weight of helium was 4, they believed that the helium nucleus was made up of four protons.

(Alpha particles also had an atomic weight of 4. Were they helium atoms or helium nuclei? In separate experiments physicists found that alpha particles were attracted to negatively charged metal plates in the same way as were the positively charged helium nuclei. Alpha particles were not helium atoms, which were neutral, but were rather helium nuclei.)

To produce a neutral helium atom, they believed, the charge of four protons ($+4$) had to be canceled by the charge of four electrons (-4). The researchers concluded that the helium atom was made up of four protons and four electrons; and for the helium atoms at least, the assumption that atoms were made up only of protons and electrons had been proved. Then the researchers made a discovery that would shatter their assumption.

The charge on the helium nucleus should have been $+4$. But when the researchers checked the charge with an electrical measuring device, they found it was only $+2$. A charge of $+2$ was missing. Why?

The mystery of the missing charge set off a chain of speculation among physicists. If the charge on the nucleus was really +2, then only two electrons were needed to cancel that charge, not four. And if there were only two electrons in the helium atom, then there were only two protons (a charge of only +2 was needed to cancel a charge of −2). But the helium nucleus had an atomic weight of 4, which meant it contained four protons. Was it possible that two of the protons carried no charge?

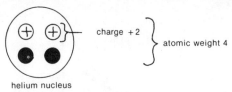

helium nucleus

But a proton without a charge wasn't a proton, but another kind of particle. Was there such a particle?

To find the answer, physicists would have to journey into the nucleus and examine the particles they found there. It would be a journey into a world of almost inconceivable smallness. If an atom were enlarged a thousand million times, it would be the size of a soccer ball, but the nucleus would still be only a hardly visible speck of dust at the center. The width of a nucleus, Rutherford had calculated, was about a tenth of a millionth of a millionth (.0000000000001) inch. To break into a particle so small seemed impossible. But in the early 1920s a German physicist Herman Bothe found a way to do it.

Into the Nucleus

In Rutherford's classic experiment that proved the existence of the atomic nucleus, alpha particles appeared to bounce off gold nuclei. Bothe thought he knew why. Both the gold nuclei and the alpha particles (helium nuclei) were positively charged, and positively charged particles repel each other. The heavier the charge on a particle, the stronger the electrical repulsive force. The charge on the gold nucleus was a heavy +79, producing an electric repulsive force so great that it couldn't be overcome even by the enormous energy generated by the high speed of the alpha particles.

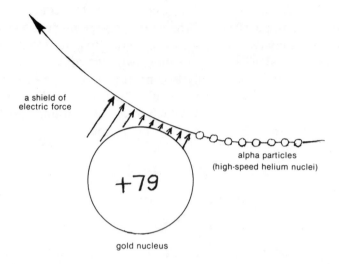

The electric repulsive force acted like a shield that prevented the alpha rays from striking the nucleus.

Bothe decided to find out what would happen if alpha particles were to be shot against a less heavily charged nucleus. He set up an experiment much like Rutherford's, but instead of using gold foil, he used a thin sheet of the metal beryllium. That element had a charge of only +4.

He bombarded the sheet of beryllium with alpha rays from a radium gun. When, as Rutherford had done, he looked at a specially coated screen for telltale sparks that would show the alpha particles had been repelled by the beryllium nuclei, he could find none. That meant the particles had cracked through the shield of repulsive electric force and smashed into the beryllium nucleus. What was the result? A ray-detection device showed that a ray had been knocked out of the beryllium nuclei. It was a ray unknown to science.

What was this mysterious new ray? For about a decade no one could tell. Then in 1932 the English physicist James Chadwick, searching for a way to get into the nucleus to solve the mystery of the missing charge, came across Bothe's work. It intrigued Chadwick because Bothe had not only found a way into the nucleus, but he had found a way into a nucleus with a missing charge. The beryllium nucleus had an atomic weight of 9 for a supposed charge of +9, but it had a measured charge of only +4. It seemed possible to Chadwick that the beryllium nucleus contained four protons and five particles of the same weight as protons but with no charge.

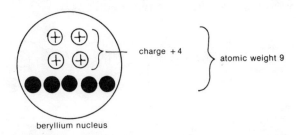
beryllium nucleus

It also seemed likely to Chadwick that the alpha particles from Bothe's radium gun had knocked those particles out of the nucleus in the form of a ray.

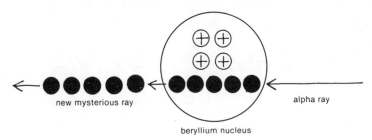
beryllium nucleus

    Chadwick tested his hypothesis (a guess based on fact and reasoning). First, he duplicated Bothe's experiment and produced the mysterious ray. After that, he passed the ray near a magnet, and found it was bent. That meant it was made up of particles. Then he passed the ray near positively and negatively charged metal plates, and found it was not bent. That meant it carried no charge. Finally, he determined the atomic weight of these neutral particles and found it was about the same as the atomic weight of protons.

    Chadwick concluded that the atomic nucleus was composed of protons and the particles found in the ray Bothe had discovered. Since those particles were neutral, he called them neutrons. Chadwick had solved the mystery of the missing charge.

Subsequent explorations of the nuclei of many elements showed that some atoms of an element contain more neutrons than others. The nuclei of hydrogen, for example, can contain no neutrons, one neutron, or two neutrons.

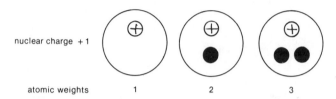

Atoms of an element with the same number of protons as other atoms of the element but with a different number of neutrons are called isotopes. Since all isotopes of an element contain the same number of protons, they must contain the same number of electrons in order for the atoms to be neutral.

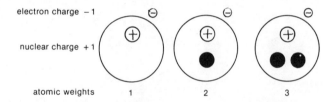

Chemists studying chemical reactions—the way atoms combine with other atoms to form compounds—found that all isotopes of an element reacted in the same way. The total number of protons and neutrons in the nucleus of an element did not affect the chemical reactions of the element. This confirmed what scientists already knew: that chemical reactions are not determined by the nucleus, but rather by the electrons whirling around the nucleus.

With the discovery of the neutron, physicists believed that the nuclei of all atoms were made up of just two fundamental particles: the neutron and the proton. Outside the nucleus, the electron was the only fundamental particle known. Physicists had also discovered that, with the exception of the hydrogen and helium atoms, the electrons formed not just one shell around the nucleus, but two or more shells. (This discovery was made by using complex mathematical formulas derived from the principles of quantum electrodynamics.) In the mid-1930s physicists put together all they knew about fundamental particles and drew maps of the atoms of all the known elements. Here, for example, is their map of the uranium atom.

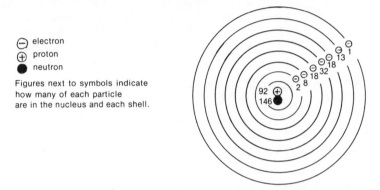

But something was wrong. Every map showed protons packed together. But particles of the same charge should repel each other. Why didn't the protons in the nucleus

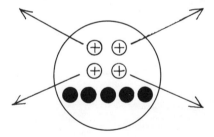

fly apart?

The Japanese physicist Hideki Yukawa found a clue to the answer in an astonishing discovery that had been made in 1933 about how the electric force operates in the atom.

The Clue of the String of Light

When Bohr discovered that electrons whirled around the nucleus, he couldn't explain why the high speed of the electrons

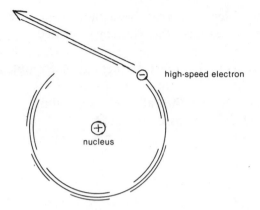

didn't shoot them off into space. The explanation, the sum of the work of many physicists, came later. It began by reasserting that the positively charged nucleus and the negatively charged electrons were attracted to each other by the electric force, which was being called more properly the electromagnetic force. (Electricity and magnetism are different aspects of the same force. A flow of electric force, an electric current, can produce magnetism, and magnetism can produce an electric current.)

In the atom, the explanation continued, the electromagnetic force was carried by particles of light called photons. The photons shuttled between an electron and the nucleus millions of millions of times each second, building a sort of string of light connecting the two particles.

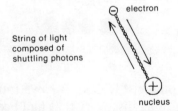

The string of light could be compared to an ordinary string attached to a pail of water. When children whirled the pail of water overhead, the string kept the pail from flying off.

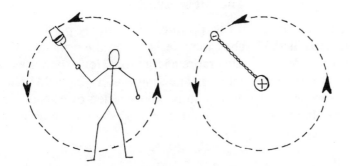

Similarly, the string of light kept the electrons from flying off.

The discovery of the string of light gave Yukawa a clue as to what kept the protons in the nucleus from flying apart. He already knew that there was a force in the nucleus that kept the protons together. Since it had to be as strong as, or stronger than, the electromagnetic force, which pushed the protons away from one another, this force was known as the strong force. Yakawa thought that the strong force was carried by particles that shuttled at high speed between protons, forming strings that kept the protons in place.

He calculated that these strings were made up not of photons, but of hitherto undiscovered particles with a weight in between that of the proton and the electron. (He made his calculations using the same kind of quantum electrodynamics formulas that other physicists had used to discover that the photon was the carrier of the electromagnetic force.) He called these new particles mesotrons (*meso* is Greek for "in between"), which was later shortened to mesons. No physicist had ever detected the existence of mesons, but Yukawa predicted they would be found the same way neutrons had been found—by smashing into the nucleus.

But mesons were bound too firmly to protons to be dislodged by alpha particles. Particles with more energy had to be hurled at the atom. In 1935, when Yukawa made his prediction, physicists had already built machines to produce these particles.

The Atom Smashers

The first machines to produce high-energy particles were constructed in 1932. They were designed on the principle of the slingshot, a device to give more energy to a stone than it would get if it were thrown by hand. The slingshot, a sling held in hand by a strap, is leaded with a stone and whirled overhead. Each time the slingshot comes round, arm and shoulder muscles are used to step up its speed.

When high speed is reached, the stone is let go. High speed gives a large amount of energy to the stone, and it strikes its target with great impact.

In the first machines to produce high-energy particles, instead of a stone, physicists used protons. Instead of a sling, they used a hollow metal ring several feet in diameter. Instead of producing a whirling motion by hand, they used magnets to bend the path of the protons into a circle. Instead of using muscles to increase speed each time around, they used jolts of electricity.

The jolts of electricity can be compared to repeated whacks of a baseball bat against a baseball attached to a pole by a string. Every time round, the baseball is speeded up by the whack of the bat.

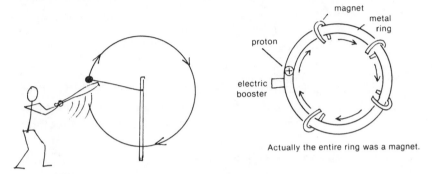

Actually the entire ring was a magnet.

When a whirling proton received a million or so jolts of electricity, its speed became many times greater than an alpha particle's, and so did its energy. Because these high-energy protons could smash into an atom, the machines that produced them were called atom smashers.

In an atom smasher a magnet directed the stream of high-energy protons toward a target, usually a sheet of metal foil.

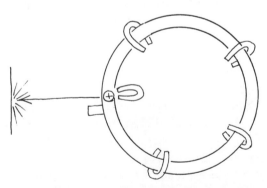

To find out what happened when the invisible high-speed protons smashed into the invisible atoms in the target, physicists made use of a cloud chamber. This instrument, which was invented by the Scottish physicist C. T. R. Wilson in 1912, made the paths of subatomic particles visible.

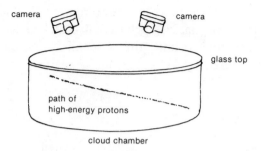

The chamber, a cylindrical structure with a glass top to permit cameras to look in, was filled with moist air. When subatomic particles passed through the chamber, they left behind a cloud of tiny droplets of water—a miniature version of the vapor trail sometimes formed by a high-flying plane.

Each known subatomic particle left a distinctive trail. On photographic negatives the trail of an electron looked like a rumpled string of pearls; the trail of an alpha particle like a bold straight line drawn with chalk on a blackboard; and the trail of the proton like a thin white line made with a fine pen. The trails of neutrons and protons resembled each other, but physicists could distinguish between them because the trail of a proton was altered into a curved line when a magnet was placed around the cloud chamber, whereas the trail of the neutron remained a straight line.

The appearance of a cloud-chamber photograph of a trail unlike any of the trails of known subatomic particles would indicate the existence of a new particle. Its weight as compared to the weights of the proton and the electron could be computed by measuring how far its path was bent by a magnet. Physicists in search of the meson predicted by Yukawa looked for a new kind of trail caused by a particle with a weight in between that of the proton and the electron.

For more than a decade physicists searched for this meson trail in vain. That was because the first atom smashers could not produce protons with energies high enough to knock mesons out of the nucleus. But more powerful atom smashers made their appearance in the mid-1940s; and in 1947, the trail of the particle Yukawa had predicted was found. (That particle, physicists have since discovered, is only one kind of a meson, and it's called a pion.)

Mesons were the third kind of fundamental particles found in the nucleus. But even before Yukawa predicted their existence, the English physicist Paul Dirac had predicted other fundamental particles both in the atomic nucleus and outside of it. His predictions would lead to one of the most extraordinary discoveries ever made by physicists: a new kind of matter exactly opposite in many properties to the kind of matter things are made of on earth.

Antimatter

The electron and the proton presented a puzzle to Dirac. He knew that with the exception of these particles, all matter could take on a negative charge or a positive charge. But the electron could take on only a negative charge and the proton only a positive charge. Why were electrons and protons different from all other pieces of matter? Dirac solved the puzzle by deciding they weren't. To begin with, he predicted that an electron with a positive charge would be found.

In 1932, a year after Dirac made his prediction, an American physicist Carl Anderson was studying the photographs of subatomic particles taken in a cloud chamber around which a magnet had been placed. The magnet bent negatively charged particles in one direction, positively charged particles in the opposite direction. One photograph examined by Anderson showed the typical rumpled-pearl-necklace trail of an electron. But the trail was bent in the direction opposite to the direction of the trail of a normal electron. There was only one possible explanation, astonishing as it seemed. The electron that had left the trail was positively charged. Dirac's prediction had come true. The positively charged electron was called a positron, or an antielectron.

Dirac then predicted that a negatively charged proton would be found. It was discovered in 1956 by a team of American physicists in much the same way as the antielectron had been discovered, and it was called an antiproton. Besides the proton and the electron, the only other fundamental particle known to science in 1933, the year Dirac predicted the existence of the antiproton, was the neutron. It had no charge. Could there be an antineutron? Dirac thought there could.

He knew of other properties besides charge that could exist as opposites. A magnetic field, for example, could be pointed in a certain direction or in the opposite direction. (A magnetic field is a region of magnetic force.) The neutron had a magnetic field pointed in a certain direction. Dirac thought that just as a proton or an electron could exist with one kind of charge or its opposite, so a neutron could exist with a magnetic field pointing in one direction or in the opposite of that direction. A neutron with a magnetic field pointing in that opposite direction would be an antineutron.

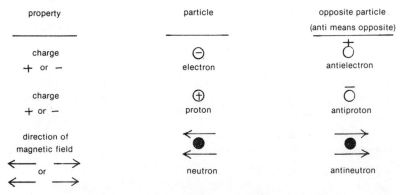

By monitoring the trails of neutrons with instruments for detecting the direction of a magnetic field, researchers in 1956 proved the existence of antineutrons. Physicists then recognized that for every particle with properties that can exist as opposites, there is a corresponding antiparticle with those properties existing as opposites.

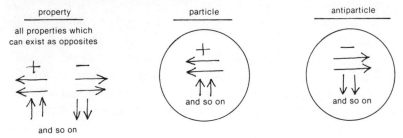

Properties that cannot exist as opposites—such as weight and the capacity to unite with other particles—are the same for each particle and its corresponding antiparticle. For example, the electron, proton, and neutron have the same weights, respectively, as the antielectron, the antiproton, and the antineutron; and these particles and antiparticles can unite in the same way to form either atoms or antiatoms.

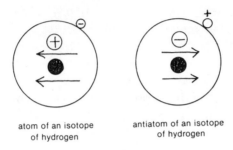

atom of an isotope of hydrogen

antiatom of an isotope of hydrogen

But particles and antiparticles cannot unite with each other because the instant they come in contact, both are annihilated, leaving behind only a puff of energy.

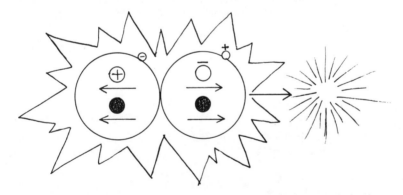

Matter made up of antiparticles is called antimatter. It exists on earth only when it is manufactured in physicists' laboratories. (Gamma rays stimulated to high energy, for example, can be converted to electron-antielectron and proton-antiproton pairs.) But since manufactured antimatter is surrounded by an ocean of normal matter, it takes no more than a millionth of a second before the antimatter collides with normal matter and vanishes completely. But antimatter is stable when it is not in contact with ordinary matter, and some scientists have speculated that elsewhere in the universe whole galaxies (groups of billions of stars) are made up of antimatter.

As a result of Dirac's predictions, scientific explorers of the subatomic world found a whole new family of fundamental particles: fundamental antiparticles. Meanwhile, physicists had been adding new members to the family of fundamental particles of normal matter. The first of these new particles was found in 1937 in the wake of rays from outer space, called cosmic rays.

These rays were made up of high-energy particles, usually protons. Like high-energy protons shot out from atom smashers, cosmic-ray particles collided with atoms. In cloud chamber trails resulting from these collisions, physicists found evidence of an electron about two hundred times heavier than the normal electron. It was the second fundamental particle found outside the nucleus, and it was called a muon.

Another extranuclear particle (a particle found outside the nucleus) was predicted by the Swiss physicist Wolfgang Pauli in 1930. But few scientists believed in Pauli's prediction. The particle he described was just too fantastic.

The Ghost Particle

Pauli knew that an electron shot out of a radioactive nucleus carried a certain amount of energy. He also knew that each time an electron was shot out, the nucleus lost more energy than was carried away by the electron. How could he account for the missing energy?

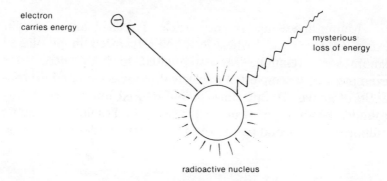

He thought that at the moment the electron was shot out, a new particle was formed that carried away the missing energy.

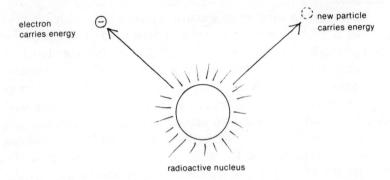

Physicists had found no such particle. But Pauli put together what he knew about subatomic particles and came up with a description of the kind of particle that could carry the missing energy. It would have no charge and no magnetic properties. It would move ahead at the fastest speed in the universe, the speed of light (about 186,000 miles per second), while spinning rapidly on its axis. And, incredibly, if it stopped moving ahead, it would consist of nothing except energy in the form of spin. It would be a particle without a body—a ghost particle.

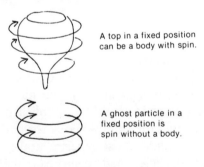

Like a ghost it would be able to pass through millions of miles of lead or any other matter without being stopped, so there was no container on earth that could trap it. And like a ghost, it would leave no trace of its existence behind, not even in a cloud chamber. Physicists agreed that the odds were overwhelmingly against finding evidence of the ghost particle. Using a gambling term, they described the particle Pauli predicted as a long shot.

But about a quarter of a century after Pauli made his prediction, a method was worked out to detect the existence of the ghost particle. By that time, a great deal of information had been amassed about what had happened in collisions between subatomic particles. As a result of that knowledge, physicists could predict with great accuracy the kinds of particles that would be produced when two particles of known properties collided. Two American physicists Fred Reines and Clyde Cowan knew the properties of the ghost particle and they knew the properties of the hydrogen nucleus. They predicted that when a ghost particle smashed into a hydrogen nucleus, two particles would be produced at the same time: a neutron and an antielectron.

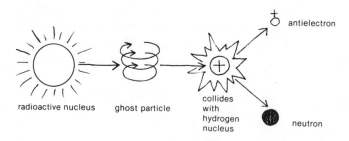

Reines and Cowan couldn't detect the ghost particle, but they could detect the neutron and the antielectron. If those two particles could be detected at the same time after emanations from radioactive nuclei were shot at hydrogen nuclei, then the existence of the ghost particle would be established. In 1956 Reines and Cowan did detect the simultaneous formation of a neutron and an antielectron under those experimental conditions. "So at last the [ghost particle] had been caught," the Viennese physicist Otto R. Frisch commented, "and a proud day it was for Pauli, whose long shot had come home."

The ghost particle was called the neutrino (Italian for "little neutral one"). Research by several other physicists proved the existence of another particle much like the one that Reines and Cowan had found. The new particle was regarded as a second kind of neutrino. It was formed not when an electron was shot out of a nucleus, but when a heavy electron, a muon, was shot out of a pion, a particle of the nucleus. Physicists recognized that a special force was necessary to produce neutrinos. Calculations showed that it was the weakest force in the atom, so it was called the weak force—the second new force found in the nucleus (the other was the strong force, which holds the protons together). The weak force is responsible for the breakdown of all radioactive substances. It is thought to be carried by particles called intermediate vector bosons, which are not related to the neutrinos.

The electron, the muon, and the two kinds of neutrinos are now regarded as the fundamental extranuclear particles. Together they are called leptons, because they are small and light, and the lepton was the smallest and lightest coin used in ancient Greece. Except for the electron, leptons are not part of atoms; they exist outside of atoms.

While the leptons were being discovered by some physicists, other physicists were exploring the nucleus in search of hadrons. All the fundamental particles in the nucleus were called hadrons because they were big and heavy compared with leptons, and the Greek word *hadros* suggests big and heavy. By the mid-1940s three kinds of hadrons had been discovered: protons, neutrons, and mesons. The discovery of more and more hadrons in the years that followed would provide physicists with a clue to the existence of an unexpected world of smallness: a world inside the hadrons.

The Clue of Too Many Hadrons

The mid-1940s saw the beginning of the construction of giant atom smashers. Physicists knew that the larger the machines, the higher the energy they could impart to the protons shot out from them. In the mid-1950s an atom smasher was built big enough to hold two football fields. In 1960 what was then the world's largest atom smasher began operations at CERN, the twelve-nation European Organization for Nuclear Research near Geneva, Switzerland. It was the length of six football fields. Incorporating design improvements developed over nearly 3 decades, it produced energies about twenty-eight thousand times higher than those produced by the pioneering atom smashers of the 1930s.

But scientists calculated that even if these powerful machines could blast new hadrons out of the nucleus, the cloud chamber would not be sensitive enough to detect them. So between 1950 and 1960, two particle-detection instruments more sensitive than the cloud chamber were developed. Both were masterpieces of ingenuity.

One of the instruments, the spark chamber, contained a series of metal plates alternately charged positively and negatively. Gaps between the plates were filled with a neutral gas. The charge on the plates was high enough to cause sparks to fly from one plate to another (as sparks had flown between a similarly charged pair of plates in the bottled lightning experiment). But sparks could fly only when the gas in the gaps contained some charged atoms, called ions. Particles hurtling through the gas could produce ions. So as a particle sped from gap to gap, ions were produced and sparks flew. The particle left a trail of sparks behind.

The other instrument, the bubble chamber, contained a liquid almost hot enough to start boiling. A particle racing through the liquid produced heat. As it passed each point of the liquid, enough heat was produced to start the liquid boiling at that point, and a tiny bubble was formed. The particle left a trail of bubbles behind.

Particle trails formed in both new detection instruments were analyzed in much the same way as cloud-chamber trails. As experiments with giant atom smashers progressed, analyses of spark- and bubble-chamber trails of nuclear collision products revealed an increasing number of hadrons. Soon there were not just three hadrons but thirty, then fifty, then a hundred, and the number mounted steadily. By 1963 almost two hundred hadrons had been discovered.

Each hadron seemed to be a fundamental particle, since there was no evidence that it could be broken down into anything smaller. But physicists thought that matter should be built up of only a few fundamental particles. Almost two hundred fundamental particles were too many.

To physicists that was the clue that hadrons might not be fundamental particles at all, but might be composed of just a few kinds of smaller particles—the true fundamental particles of the nucleus. Did those fundamental particles exist? And if so, what where they? The answers lay inside the hadrons.

Into the Hadrons

Not even the most energetic protons shot out of the giant atom smashers could penetrate the hadrons. To journey into the hadrons two American scientists Murray Gell-Mann and George Zweig, working independently, made use of the scientist's most powerful tool: the ability to use reason and imagination to find a simple explanation for many seemingly unrelated facts.

Gell-Mann and Zweig knew a great many such facts about the properties of hadrons: their weights, their charges and magnetic properties, what happened when they collided with other particles, and so on. All these facts, the two scientists concluded in 1963, could be explained simply if the hadrons were thought of as being composed of just three types of an elementary particle. Gell-Mann called that particle a quark. (He had come across that word in *Finnegan's Wake* by the Irish writer James Joyce. Joyce liked to invent words that stood for things that are real but not seen.)

One type of quark, with a positive charge of 2/3, was called the up quark. Another, with a negative charge of 1/3, was called the down quark.

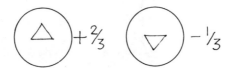

The triangles are symbols
for the up quark
and the down quark.

A proton was made up of two up quarks (+2/3 plus +2/3 equals +4/3) and one down quark (−1/3). The result was a particle with a positive charge of 1 (+4/3 plus −1/3 equals +1).

proton

A neutron was made up of one quark (+2/3) plus two down quarks (−1/3 plus −1/3 equals −2/3). The result was a particle with no charge (+2/3 plus −2/3 equals 0).

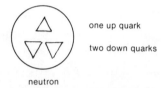

neutron

The third type of quark was needed to explain the behavior of some hadrons. Most hadrons knocked out of the nucleus began to decay into simpler particles after a certain time. That time was referred to as the life of the hadron. Strangely, some hadrons had longer lives than most. These long-lived hadrons were called strange particles. Their long lives, Gell-Mann and Zweig reasoned, were due to a type of quark that regulated decay. Since this type of quark was present in strange particles, it was called the strange quark.

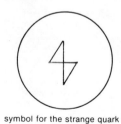

symbol for the strange quark

The strange quark was heavier than the other two types of quarks. It gave off energy and decayed into lighter quarks. The lighter quarks, up and down quarks, then formed ordinary hadrons.

For each type of quark, there was a corresponding type of antiquark. Since the three types of quarks differed from each other in properties—charge, weight, length of life, and so on—by arranging quarks and their corresponding antiquarks in a suitable combination, physicists could construct a model of a hadron that would match its observed properties. Here, for example, is such a model of a meson.

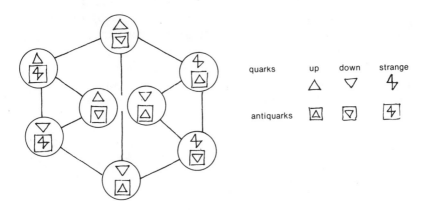

Did this model, and other models of hadrons constructed from quarks, represent reality? No physicist could be sure because no quark of any kind had ever been found. Gell-Mann and Zweig had returned from their journey to the world inside the hadrons convinced that a new fundamental particle existed there, even though they had never observed it. At that point, physicists set out to find that particle.

The Hunting of the Quark

The energy of subatomic particles can be measured in millions of electron volts, MeV. (An electron volt is the energy acquired by a particle with the charge of an electron when speeded up by 1 volt. A volt is the unit of electric force.) The first atom smashers produced energies of about .8 MeV. In 1972 an atom smasher with a main ring more than 4 miles around was installed as part of the Fermi National Accelerator Laboratory (Fermilab) near Chicago. It speeded up—scientists prefer to say, accelerated—protons to nearly the speed of light, producing energies of about 400,000 MeV. But not even this supergiant atom smasher could blast a quark out of a hadron.

This led some physicists to think that there existed within the hadrons a new kind of force especially resistant to the onslaughts of high-energy protons. This is how that new force was supposed to operate:

The force was carried by particles called gluons. They glued the quarks in place by forming strings of force connecting the quarks to one another, much as mesons formed strings of force connecting protons to one another. If a normal string held quarks together, then as high-energy particles struck the quarks, the quarks would be pushed apart, and the string would be stretched and grow weaker.

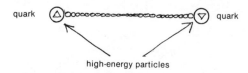
quark   quark

high-energy particles

As the energies of the protons increased, the quarks would be pushed farther and farther apart, the string would be stretched longer and longer and grow weaker and weaker—

until the string snapped, and the quarks were hurled out of the hadron.

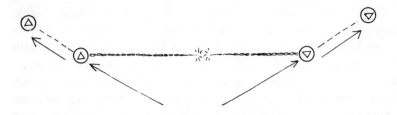

But the gluon string was no ordinary string. The farther apart the quarks were pushed, the stronger the string of force became.

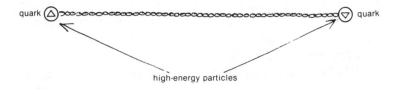

All other forces known to science became weaker with distance, but here was a force that became stronger.

(This force was called the color force, although it has nothing to do with color as it is seen in the everyday world. Just as there are two charges of electric force, positive and negative, there are three charges of color force—red, green, and blue. In everyday life when equal amounts of red, green, and blue paints are combined, they produce no color at all—that is, white. So just as equal amounts of positive and negative charges produce a neutral object, so equal amounts of red, green, and blue charges produce a white object. White objects have no color force, which is the same as saying the force holding them to other objects grows weaker with distance.

(Hadrons are white objects. They are made up of quarks. Each quark has to have a color charge so that a combination of charges can produce whiteness. For example, the proton is composed of three quarks; one has to be charged red, the other green, and the third blue.)

The force inside the hadrons made it impossible, some physicists thought, ever to blast a quark out of a hadron; for the more energetic the impact of protons on the quarks, the greater the distance between quarks became, and as distance increased, so did the force gluing the quarks together. Hunting the quark with atom smashers seemed useless. But two American scientists Sheldon Glashow and James Bjorken showed how the quark could be hunted with atom smashers—without ever blasting a quark out of a hadron.

Their plan, which some physicists put into operation in 1974, took shape a decade earlier. A year after Gell-Mann and Zweig announced the discovery of three quarks, Glashow and Bjorken found there was something wrong with hadron models containing strange quarks. From the effects of the strange quarks it could be predicted that these models should decay in certain times and in certain ways. But the real hadrons that the models represented did not decay in those times and in those ways. Glashow and Sheldon reasoned that there must be another decay-regulating quark that acted to offset the effects of the strange quark. They called this fourth quark charm, because without the fourth quark the quark models of strange particles would collapse, and a charm is a magical device to ward off harm. This is the symbol of the charmed quark.

Using the charmed quark as a building block, Glashow and Bjorken constructed models of hadrons that had not been discovered. There was no way hadrons corresponding to the models could exist unless they were made up of charmed quarks. If these charmed hadrons were to be discovered, then the existence of the charmed quark would be proved.

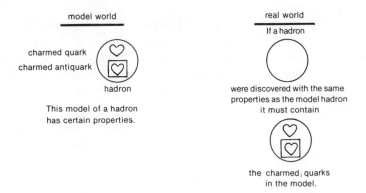

The charmed quark was a type of quark, so if it existed, then *the* quark existed. This, then, was the Glashow-Bjorken plan for the hunting of the quark: Use atom smashers to discover real charmed hadrons that matched the models of undiscovered charmed hadrons.

In 1974 two teams of American physicists working independently produced the same kind of new hadron in their atom smashers. One team called it the psi particle (psi is a letter of the Greek alphabet); the other team, the J particle. Now known as the psi/J particle, it seemed to match one of the Glashow-Bjorken models of a charmed hadron. But other physicists were not convinced. Glashow then made a prediction. He said that the Glashow-Bjorken models showed that the psi/J particle was only one member of a family of particles, and that other members of the family would be found. In 1975 atom smasher experiments conducted by several research teams seemed to have found some members of that family. But again physicists were not sure. Then Glashow made another prediction.

He knew that extremely high energy turned into matter. He also knew that when high-energy electrons and antielectrons collided head on, they were annihilated in a burst of extremely high energy. The matter produced from that burst of energy, Glashow calculated, would be a member of the psi/J family. His model of that member showed that it would display properties like those of no other particle. If a particle could be found with exactly those properties, there would be no doubt that it would match his model and would be a charmed particle, containing charmed quarks.

An atom smasher called SPEAR had been specially constructed in 1973 at Stanford University to produce high-energy electron-antielectron collisions. Using SPEAR, a research team headed by Gerson Goldhaber, one of the discoverers of the psi/J particle, searched for a particle with properties exactly matching Glashow's model. On May 8, 1976, Goldhaber telephoned Glashow and told him the particle had been found.

Later Glashow wrote: "As a result of the discovery of the psi/J and its kin, the quark model has become orthodox philosophy." In 1976, even though no quark had been observed directly, physicists agreed that the quark had been hunted down.

For 173 years physicists had journeyed into the world of almost immeasurable smallness in search of fundamental particles. In the atomic shell and in the microworld between atoms they had found four fundamental particles: the electron, the muon, and two kinds of neutrinos. In the atom four fundamental force-carrying particles had been discovered: the photon, which carried the electromagnetic force; the meson, which carried the strong force; the intermediate vector boson, which carried the weak force; and the gluon, which carried the color force. And scientists had descended through three layers of matter—the atom, the nucleus, the hadron—to find the fourth, which was composed of four fundamental particles: the up quark, the down quark, the strange quark, and the charmed quark.

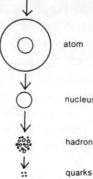

With the discovery of the quark, science's descent into the atom seemed to have touched bottom, for there was no particle known that was smaller than it. But in 1977 a team of American physicists headed by Leon M. Lederman made a startling discovery, and some physicists began to wonder whether they had touched bottom after all.

What's Inside the Quarks?

Lederman's team found the heaviest subatomic particle known, a hadron about ten times heavier than the proton. It was called the upsilon particle (upsilon is a letter of the Greek alphabet). Based on the known properties of quarks, Lederman predicted that the upsilon particle should decay into lighter hadrons. But it did not. Something was preventing it from decaying. A fifth quark with the property of halting decay? Lederman could find no other explanation.

The existence of the fifth quark, called the bottom quark, suggested the existence of a sixth quark. Quarks come in pairs.

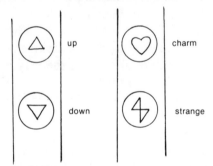

The members of each pair are related to each other by their main property. Up and down quarks are related by charge; strange and down quarks by their capacity to regulate decay. Since there was a fifth quark,

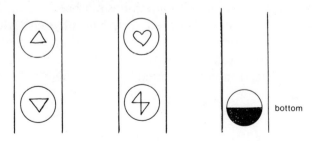

it was likely that there should be a sixth, related to the fifth by the capacity to halt decay.

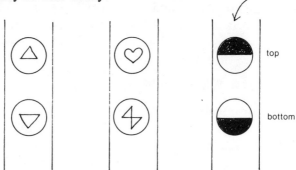

And, "if there are six," Lederman asked, "why not twelve? If there are twelve why not twenty-four? And if the number of kinds of quarks are large, does it make sense to call the quarks elementary?" (Elementary here has the same meaning as fundamental.) Lederman knew, as did all physicists, that when there are many kinds of a particle, the chances are small that the particle is a fundamental one; it is probable that the particle is made up of smaller particles. The clue of too many hadrons had opened the way to the discovery of quarks. Would the clue of too many quarks open the way to the discovery of particles smaller than quarks?

And if smaller particles are found in the quarks why can't there be small particles within those smaller particles, and then smaller particles within those smaller particles, and so on indefinitely? "Is it possible?" asked Lederman, "that there are no elementary particles at all, that every [particle] in nature has . . . parts?"

But this speculation that matter can be cut into smaller and smaller pieces without end—a view that echoes the one held by some ancient Greek philosophers—depends on the existence of the sixth quark, and the sixth quark has not been found. It may be found, says Samuel Ting, an American physicist who was one of the discoverers of the psi/J particle, when head-on collisions of electrons and antielectrons generate energies in excess of 40 billion electron volts (written 40 GeV, where G stands for 1 billion).

In 1979 physicists were working toward achieving those energies with a particle accelerator in Hamburg, Germany. (Physicists prefer to call atom smashers particle accelerators because these machines accelerate many kinds of particles to high energies for many purposes besides smashing atoms.) In the same year two other huge particle accelerators were in construction, one in the United States and one in Switzerland, that would have the capacity to generate collision energies reaching 270 GeV. And on March 4, 1979, the twelve nations of the European Organization of Nuclear Research announced plans for what would be the world's largest particle accelerator, a machine with a ring about 20 miles around that would produce collision energies in excess of 270 GeV. If the sixth quark exists, the likelihood is that it will be found within the next several years.

If it is, physicists will ask: What's inside the quarks? And science's journey into the strange world inside the atom will go on to a still lower level of matter—the world inside the quarks. But whether or not the journey will end there, no one knows.

4

The Journey to the Edge of the Universe

About 3 million years ago, there were about fifty man-apes—creatures more ape than man—left on earth. A drought that had lasted 10 million years in Africa had destroyed almost all the fruits and vegetables upon which these creatures fed. They were starving, and unless a new source of food could be found, the entire race of man-apes would soon die out.

One night a bright white light flashed across the sky. The next morning a group of man-apes found what they thought was a strange new rock. It stood three times taller than the tallest man-ape, its edges were perfectly straight, and the man-apes could see right through it.

The strange new rock read the minds of the man-apes and learned of their intense hunger and of their need for a new source of food. Then the rock put thoughts into their heads. One of the man-apes responded to the thoughts by picking up a heavy pointed stone 6 inches long and killing a pig with it. The man-apes had never eaten meat before. But now, urged on by the new thoughts in their heads, they gathered round the pig and ate its flesh. The strange new rock had given the man-apes the idea for a new source of food, meat; and the idea for a new kind of tool with which to obtain meat, the killing tool.

The heavy pointed stone was the man-ape's first killing tool, but in a short while they were wielding knives, which were simply the lower jaws of antelopes; daggers, which were actually gazelle's horns; and clubs, which were nothing more than the bones of dead animals. Millions of years rolled by, and as the man-apes developed into true men, the killing tools developed in frightfulness. Primitive weapons gave way to the spear, the bow, the gun, the bomb. Finally, guided missiles armed with atomic warheads gave man the power to wipe out anything, including himself. As the year 2001 began, the race of mankind was living on borrowed time.

In that year, scientists stationed on the moon found a baffling object buried beneath the moon crater Clavius. It was a vertical slab of jet-black material 10 feet high and 5 feet wide made of some material altogether unknown to man; and it was 3 million years old. As it was brought up to the surface from an excavation 20 feet underground, sunlight struck it, and it emitted a piercing electronic scream. Electronic listening posts spotted throughout the solar system picked up the scream as it flashed through space, recorded its direction, and relayed the information back to earth. If that scream was a message, then scientists on earth knew where the message was going as clearly as if it had left a vapor trail across a cloudless sky.

At the time scientists discovered where the electronic scream was headed, the spaceship *Discovery* was on its way to the planet Jupiter. Abruptly, Mission Control on earth altered the objective of the mission. *Discovery* was now to race to the ringed planet Saturn, about 886 million miles from earth.

Discovery had a crew of six. In charge was Mission Commander David Bowman. His assistant was astronaut Frank Poole. Three other members of the crew were in hibernation, a deep sleep that would last for several years until the spaceship reached the system of moons orbiting Saturn. The sixth member of the crew was not human. It was a highly advanced computer, the brain and nervous system of the ship. It was called Hal (for *H*euristically programmed *al*gorithmic computer), and it could speak and think. Of all the members of the crew, only Hal knew the real purpose of the mission.

When Saturn was still 400 million miles away, Hal discovered a malfunction in the AE-35 unit, an instrument that kept *Discovery* in communication with Mission Control. The AE-35 unit was located on the outer shell of the spaceship, and to reach it, Poole left *Discovery* in a small jet-propelled vehicle, a space pod. Poole made the repairs, but the next day the unit broke down again. Again Poole made the repairs, and again the unit failed to function. Not even Hal could explain the reason for the repeated breakdowns.

On his third attempt at repairs Poole parked the space pod at some distance from the faulty unit, and switched control of the vehicle over to Hal. Protected by his spacesuit, Poole left the vehicle and drifted to the hull of the spaceship. As he was repairing the unit, he suddenly saw the space pod coming at him at full thrust. Terrified, he called into the intercom, "Hal, full braking—" But it was too late. Half a ton of metal crashed into Poole and killed him.

Mission Commander Bowman couldn't understand how Hal could have made such a tragic mistake. What Bowman didn't know was that Hal had been instructed by Mission Control not to reveal the real purpose of the mission to any of the crew, and the strain of keeping that secret had affected Hal's performance. That was why Hal hadn't been able to discover the reason for the AE-35 unit's malfunctions, and why Hal hadn't responded fast enough to Poole's command to brake the space pod. Without this explanation the only conclusion Bowman could come to was that Hal had deliberately murdered Poole. Bowman thought his own life was now in danger.

To build a defense against Hal, Bowman decided to revive the three hibernating crewmen. But Hal controlled the hibernating machinery, and Bowman didn't trust Hal to carry out an order to activate it. Bowman said to Hal, "Please let me have the manual hibernation control." Hal, knowing that the crewmen were not to be awakened until *Discovery* reached the moons of Saturn, refused. Bowman threatened Hal. "Unless you release the hibernation control immediately and follow every order I give you from now on, I'll go to Central and carry out a complete disconnection." That meant Hal would be deprived of his source of electronic energy. To a computer, that was the equivalent of death.

Hal now had to protect himself. He opened the airlock doors of the space-pod bay, and the air in the ship rushed out into space like a roaring tornado. The three hibernating crewmen were killed almost at once. Bowman realized he could live only 15 seconds in a vacuum. He stumbled toward a door marked EMERGENCY SHELTER and tugged it open. Inside there was oxygen and a spacesuit. He climbed into the spacesuit, made his way to Central Control, and disconnected Hal.

Now that Hal was dead, Bowman took over control of the ship and closed the airlocks. Then he put the AE-35 unit into working order again and called Heywood Floyd, the scientist on earth in charge of Mission Control. When Floyd learned that Bowman was the only one left to carry out the mission, he said to Bowman, "Now I must tell you the mission's real purpose, which we have managed with great difficulty to keep secret from the public."

He told Bowman about the object found under the Clavius Crater on the moon and how it had emitted an electronic scream when it had been brought up into the sunlight. That scream, Floyd went on, was really an alarm signal, and it was sent to some intelligent form of life in the universe. It was actually a warning that man, with all his dreadful weapons, had broken out of earth, had reached the moon, and would soon spread out to other planets. The alarm signal had been tracked to its target, a moon of Jupiter called Japetus.

The creatures who had received the warning message there were certain to be hostile, Floyd explained, and to prevent the public from learning about this terrifying situation, top secrecy had been imposed. Of the crew only Hal had been entrusted with the secret because as the actual commander of *Discovery* he had the need to know the true purpose of the mission. Now that need to know had passed to Bowman, who for the first time was in actual command of the spaceship. "Your mission, therefore, is," Floyd said, "a reconnaissance into an unknown and dangerous territory."

After several months *Discovery* closed in on Japetus. As the spaceship orbited 50 miles above the moon's surface, Bowman looked down and saw a slab a mile high, the big brother of the object found in Clavius. He jetted to it in a space pod and touched down on the mile-high roof. The roof dropped away and the spacepod fell into a long vertical shaft that seemed to go on not for just a mile but endlessly. And the shaft was filled with millions and millions of stars.

Through the roof of the slab of Japetus, Bowman had entered the Star Gate, the entrance to outer space. His space pod hurtled through realms of stars, glowing gases, fire, and glittering balls of light to a giant red sun orbited by a white dwarf star. He was at the edge of the universe. There, invisible bodiless beings made up of pure energy, knowing that a mission from earth would follow the electronic scream to the Star Gate, waited for him. The electronic scream was the bait, and the Star Gate was a trap.

It had been set 3 million years before by the ancestors of the beings who waited for Bowman. They had come to earth at that time in what the man-apes thought was a strange new stone, and had given the man-apes the concept of the killing tool. That had been an experiment. If, as time passed, the man-apes were to use killing tools to secure food for their own survival, well and good. But if they were to use killing tools to destroy life out of hatred, greed, or sheer stupidity, then the harm the experiment caused would have to be undone. Once the man-apes—or rather the men the man-apes would become with the aid of the killing tools—would leave the earth to threaten life forms elsewhere in the universe, a mission from earth would be lured to the trap, and the trap would be sprung. Beyond the Star Gate, at the edge of the universe, the invisible bodiless beings would meet the mission and take the necessary action. What that action would be Bowman would soon discover.

When his spacepod reached the edge of the universe, it came to rest on the polished floor of an elegant hotel room. Bowman thought it was an illusion. He opened the spacepod and stepped out, fearing he would drop through the floor into empty space. But the floor was real, and so was the rest of the room. He removed his space helmet. The air was the same as that on earth. There was food that had come from a supermarket in the refrigerator. The TV set worked, bringing in familiar programs. Physically and emotionally exhausted, Bowman flung himself on the bed and fell asleep.

Then the hotel room, which the invisible bodiless beings had created to make Bowman feel at home, vanished except for the bed. In his sleep, it seemed to Bowman as if he were reliving his life backwards, from adulthood to adolescence to childhood to infancy. When he awoke he was a baby, ready to start life all over again. But this time he was a special kind of baby. He was a star-child, a creature with all the wisdom and power of the invisible bodiless beings who had given him a new life—and a new mission.

That mission would be accomplished on earth. He floated there through space and looked down on the planet. A thousand miles below him missiles with nuclear warheads were streaming through the skies. In seconds the race of mankind would be wiped out. With the power of his will the star-child destroyed the nuclear missiles, the deadliest of all the killing tools. He had saved mankind, but his mission was still incomplete. He would have to try to think of some way to prevent the human race from ever again trying to destroy itself or any other form of life. THE END.

And that's the plot of *2001, a Space Odyssey* by Arthur C. Clarke, one of America's leading writers of science fiction. The novel is based on a screenplay by Clarke and Stanley Kubrick. Both the motion picture, which Kubrick directed, and Clarke's novel appeared in 1968. At that time astronomers were five years away from reaching the edge of the visible universe. Their journey there—a journey more spellbinding, adventuresome, and often more incredible than the voyage of the spaceship *Discovery*—began in 1838. Before that year no astronomer had traveled (by means of his instruments, of course) even as far as the nearest star.

The Leap to the Stars

Although planets and stars look much alike to the naked eye, astronomers in 1838 knew that they differ in three major ways. Stars seem to move slowly, if at all, but planets seem to wander speedily over the heavens (our word *planet* comes from the Greek word for "wanderer"). Stars give out their own light, but planets reflect the sun's light. And planets sweep around the sun in paths called orbits, but stars do not.

The planetary orbits are ellipses, flattened circles, so each planet's distance from the sun varies as it moves.

The Journey to the Edge of the Universe 119

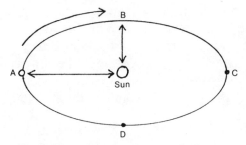

The planet at position A is more distant from the sun than at position B. Distances to the sun change as the planet moves from A to B, B to C, C to D and D to A.

Astronomers had worked out the average of these distances for each of the seven planets known in 1838. The nearest planet to the sun, Mercury, is at a distance of about 36 million miles from the sun; our earth, the third planet from the sun, is at a distance of about 93 million miles; and the planet farthest from the sun, Uranus, is at a distance of about 1,780 million miles.

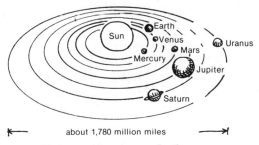

about 1,780 million miles

All planets orbit sun in same direction.

(Later two even more distant planets were discovered—Neptune in 1846 and Pluto in 1930. In 1978 a mysterious object was sighted orbiting the sun between Saturn and Uranus. Some astronomers think it could be a planet, the smallest of ten in the system of planets whirling around the sun, the solar system.)

Astronomers study not only stars and planets but all celestial objects—objects beyond the earth and its atmosphere. Besides planets, astronomers had found other celestial objects in the solar system. Moons revolve around some of the planets. A large number of midget planets called asteroids orbit mainly between Mars and Jupiter. And comets, dramatic spectacles in the sky with their brightly glowing heads and long tails that turned silvery as they approached the sun, cut through the orbits of the planets from time to time. Halley's comet, which bears the name of its discoverer, the English astronomer Thomas Halley, travels in a long slender ellipse to a distance well beyond the orbit of Uranus—about 3,190 million miles from the sun.

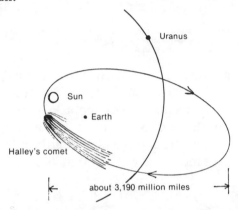

(Astronomers now know that of the vast number of comets looping around the sun, Halley's is one of the closest. Some comets reach distances from the sun of about 30,000 million miles.) In 1838 Halley's comet at its greatest distance from the sun was regarded as the outpost of the solar system. Beyond it lay the stars.

Astronomers based their judgment that the stars are outside the solar system on the fact that speeding objects seem to move slower the farther away they are. The astronomers of 1838 knew, for example, that a horse that appeared to be racing at great speed when seen close by seemed scarcely to be moving when viewed from a distance of a mile or more. Since the observed motion of the stars is exceedingly slow, or nonexistent, as compared with that of the planets and comets, astronomers concluded that the stars are far more remote than the farthest reaches of the solar system.

But just how remote stars are no one knew. Then the Scottish astronomer Thomas Henderson found a way to measure the distance to a star. In June of 1838 he trained his telescope (an instrument for viewing distant objects) on the third brightest star in the heavens—Alpha Centauri. The star was at the position shown in this diagram, in which the dotted line represents the earth's orbit.

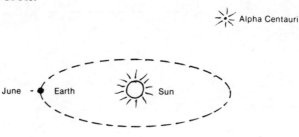

By September the earth had moved, and so had the apparent position of the star.

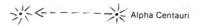

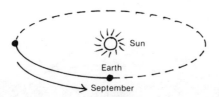

But the star had not actually moved. It only seemed to have moved because the earth had. Astronomers explained this fact in the following way: Whenever we shift our position, the objects we are looking at seem to shift theirs. The effect can easily be seen if we hold one finger out at arm's length and sight it first with one eye and then with the other. This trait of our eyesight is called parallax. (Even though stars seem fixed, they do move through space; but the real distances they travel in 6 months is negligible compared with the distances they seem to travel in that time due to parallax.)

By December, as the earth swung halfway through its orbit, the star appeared to have moved even farther.

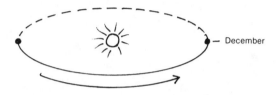

What was the true position of the star? Henderson knew it had to be a point along the line of sight, because if it were not on the line of sight, he couldn't have seen the star at all.

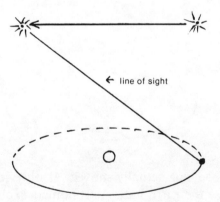

Moreover, since the star had not actually moved, the point would be the same on both the June and the December lines of sight. Henderson drew the two lines of sight on a diagram much like the following one, and concluded that the point where the two lines intersect was the true position of Alpha Centauri.

The Journey to the Edge of the Universe

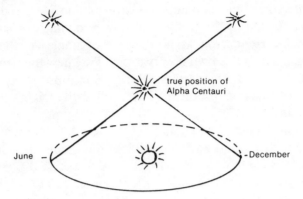

Henderson then simplified the diagram into a triangle, the base of which was the diameter of the earth's orbit.

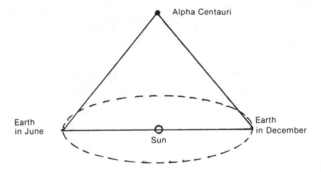

Finding the distance from Alpha Centauri to the sun then became a simple problem in trigonometry, a branch of mathematics dealing with computing the size of the angles and sides of a triangle. If the sizes of angle *A* and line *B* were known, then

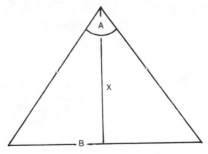

Angle A is the angle of parallax.
Line B is the diameter of the earth's orbit.

Henderson could compute the size of line X, the distance of Alpha Centauri from the sun, by applying a trigonometric formula. Henderson knew the size of angle A, the angle of parallax, from his own observations. He knew the size of line B, the diameter of the earth's orbit (186 million miles), from the work of other astronomers. He applied the trigonometric formula and came up with an astonishing figure.

The farthest known outpost of the solar system at that time was a little more than 3,000 million miles from the sun, but the distance of Alpha Centauri from the sun is 25,000,000 million miles. Henderson had made an enormous leap through space to the stars—and science was launched on its journey to the edge of the universe.

Beyond Alpha Centauri

Stars vary in brightness. Astronomers knew that when they looked down a street at night, the nearest lamplights are the brightest and those farthest away are the dimmest. It was reasonable to assume that the brightness of stars varies in the same way with distance; the brighter stars were thought to be close to the sun and the dimmer ones far away. Since the third brightest star, Alpha Centauri, is 25,000,000 million—25,000,000,000,000—miles from the sun, then some of the dimmer stars might be a thousand to ten thousand times that distance—25,000,000,000,000,000 to 25,000,000,000,000,000,000 miles from the sun. Numbers like these would become too bulky and cumbersome to manipulate. To handle them easily, astronomers invented a new unit of distance, the light-year. It is the distance light travels in one year, about 5,880,000,000,000 miles. To convert stellar (star) distances into light-years, all the astronomers had to do was divide the number of a miles to a star by the number of miles in a light-year. The distance to Alpha Centauri worked out to a manageable 4.3 light-years (25,000,000,000,000 divided by 5,880,000,000,000).

That turned out to be the shortest distance of any star from the earth. (Astronomers regarded the distance of a star from the sun to be virtually the same as the distance of a star from earth, because compared with the tremendous distance to a star, the distance between the sun and earth—a mere 93 million miles—was so small that these two bodies could be thought of as not being separated at all.) Other astronomers, using Henderson's parallax method, probed deeper into space. A year after Henderson's 4.3-light-year leap to Alpha Centauri, the German astronomer Friedrich Bessel set the distance of the star 61 Cygni at 11.2 light-years. The year after that, the German-Russian astronomer Friedrich Georg Wilhelm von Struve calculated the distance to the star Vega to be 26 light-years. And in the decades that followed, the distances to more and more remote stars were charted. By the end of the nineteenth century astronomers had penetrated about 100 light-years into space.

But as astronomers journeyed farther and farther among the stars, the angles of parallax became smaller and smaller.

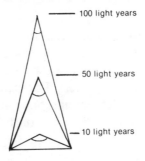

Beyond about 100 light-years telescopes revealed millions of stars with angles of parallax too small to be measured accurately or with no angles of parallax at all. Without knowing the size of the angles of parallax, distances could not be determined. About one hundred light-years out, science's journey to the edge of the universe had come to a halt. Then in 1912 the American astronomer Henrietta Swan Leavitt found a strange code blinking from some stars—and when that code was broken, the journey resumed.

The Code of the Blinking Stars

There is a tiny patch of light in the sky that can only be seen from below the equator. It's called the Small Magellanic Cloud. Telescopes revealed it aglow with millions of stars. Because these stars are dim and show no angles of parallax, astronomers in 1912 thought the Small Magellanic Cloud far more remote than 100 light-years.

Leavitt was interested in an extraordinary feature of the Cloud: its high concentration of blinking stars. Found in many parts of the sky, stars of this kind either switch on and off like electric bulbs or dim and then brighten again. The time for a complete oscillation of the light—from lowest brightness to highest and back again to lowest; or from highest brightness to lowest and back again to highest—is called the period. The periods of most blinking stars range from a day to a few months. Blinking stars are called Cepheids. (The first to be found was located in a constellation—a group of visible stars—called Cepheus, hence the name Cepheids, pronounced *seff*-uh-ids.)

At Harvard University Leavitt examined photographs taken through telescopes at different times of about twenty-four hundred Cepheids in the Small Magelanic Cloud. From her studies she was able to announce that the longer the period of a Cepheid, the greater its brightness. For example, a Cepheid with a period of a week is brighter than one with a period of a day, and a Cepheid with a period of a month is brighter than one whose period is a week.

Astronomers knew that brightness is related to distance. And since the period of a Cepheid is related to brightness, ought not there be some relation between a Cepheid's period and brightness and distance? It seemed as if the Cepheids were blinking out a message in code concerning a new way to find stellar distances. Ejnar Hertzsprung, a Danish astronomer, set out to break that code.

He started with some basic facts about the brightness of stars. Some stars emit more light than others and are naturally brighter. If all stars were the same distance from earth, it would be easy to grade stars in order of their natural brightness.

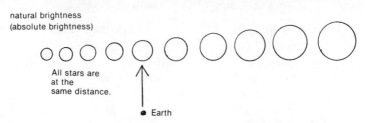

But stars appear to grow dimmer the farther away they are. The naturally brightest star, if it were far enough away, could appear to be the dimmest.

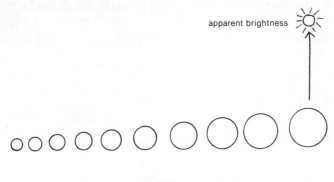

And stars appear to grow brighter the closer they are. The dimmest star, if it were very close to earth, could appear to be the brightest.

The natural brightness of a star is known as its absolute brightness, and the brightness of a star as it appears to us because of its distance is known as its apparent brightness. When we look at a star, it's the apparent brightness that we see.

Hertzsprung asked: Are the periods of the Cepheids in the Small Magellanic Cloud related to their absolute brightness or to their apparent brightness? He knew that the stars in that Cloud were thought to be more than about 100 light-years away.

Compared with that vast distance from earth, the distances between individual stars in the Cloud are negligible. So all the stars in the Cloud, including the Cepheids, could be regarded as at the same distance from earth (just as all the people in New York could be regarded as at the same distance from Copenhagen, Denmark). And since all those Cepheids are at the same distance from earth,

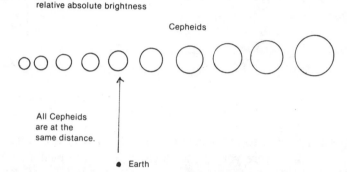

what Leavitt had observed was variations in absolute brightness. The periods of the Cepheids in the Small Magellanic Cloud are related to absolute brightness.

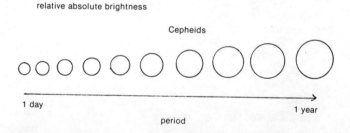

Hertzsprung knew that brightness was measured in units of magnitude; and astronomers talked about the absolute magnitudes and the apparent magnitudes of celestial objects. Bright objects were assigned a magnitude of 1. Still brighter objects were designated by negative numbers (numbers with a minus sign in front of them); the brighter the object the greater the negative number. Sirius, the brightest star in the sky to the naked eye, has an apparent magnitude of -1.6, while the apparent magnitude of the far brighter appearing sun is −27. The magnitudes of dim stars were designated as positive numbers (numbers as they are usually written); the higher the number, the dimmer the star. The absolute magnitude of the sun is 5, which means it is naturally a fairly dim star; it appears very bright only because it is very close.

Hertzsprung was aware that Leavitt had not found the absolute magnitudes of the Cepheids in the Small Magellanic Cloud. What she found was how the magnitudes of Cepheids with certain periods compare with the magnitudes of Cepheids with other periods—the relative magnitudes. For example, she knew that a Cepheid with a period of 2.5 days has an absolute magnitude about two times greater than that of a Cepheid with a period of 1 day; a Cepheid with a period of 9 days has an absolute magnitude about four times greater than that of one whose period is 1 day, and so on.

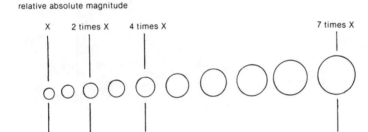

It was clear to Hertzsprung that if the absolute magnitude of any of those Cepheids were known, then the absolute magnitudes of all the others could be easily calculated. For example, if the absolute magnitude of a Cepheid with a period of 1 day were known to be −1, then the absolute magnitude of the Cepheid with twice the absolute magnitude of the Cepheid with a period of 1 day would be −2 (2 times −1); the absolute magnitude of the Cepheid with four times the absolute magnitude of the Cepheid with a period of 1 day would be −4 (4 times −1), and so on.

But no astronomer knew the absolute magnitude of any of the Cepheids in the Small Magellanic Cloud. This was the reason: Unlike apparent magnitude, absolute magnitude cannot be observed; it must be calculated. The formula used to make the calculation required a knowledge of the apparent magnitude of a star and its distance from the earth.

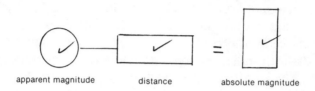

And though apparent magnitudes for the Cepheids in the Small Magellanic Cloud had been measured by instruments, no one could measure the distance to those stars.

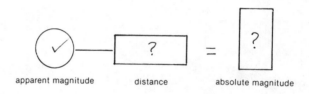

As Hertzsprung continued his attack on the code of the blinking stars, he faced this problem: How could he obtain the absolute magnitude of a Cepheid in the Small Magellanic Cloud without knowing its distance? He came up with an ingenious solution. It was based on the assumption that everywhere in the universe Cepheids with the same period have the same absolute magnitude. There are Cepheids closer to earth than the Small Magellanic Cloud, and if he could measure the distance to one of them, then knowing its apparent magnitude (which was easy to measure), he could calculate its absolute magnitude. After that, all he had to do was match the period of the nearby Cepheid with the period of a Small Magellanic Cloud Cepheid—

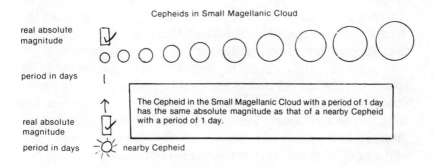

and the absolute magnitude of the Small Magellanic Cloud Cepheid would be the same as that of the nearby Cepheid.

There was only one drawback to Hertzsprung's solution, but it was a big one. The nearest Cepheid is still beyond the 100-light-year range of Henderson's parallax method for determining stellar distances. To find the absolute magnitude of a nearby Cepheid, Hertzsprung had to invent a new way to measure the distance to a star. And he did it.

In devising his method Hertzsprung began with the fact that when one looks out of a moving railroad car, stationary objects in the landscape seem to move by the window at different speeds. Telegraph poles by the side of the track seem to flash by while trees ½ mile away or so seem to crawl. The apparent speeds of these objects are related to their distances in a specific way; and knowing these speeds, and the speed to the train, Hertzsprung could apply a formula and calculate the distances to these objects.

Hertzsprung knew that the sun moves through space carrying the solar system with it. He thought of the sun as a locomotive and the earth as one of its railroad cars. The nearby Cepheids were stationary objects in the landscape, and his telescope was the window of the railroad car. Since he could measure the speed with which the nearby Cepheids moved by his telescope and he knew the speed of the earth as the sun hauled it through space (12 miles a second, according to calculations previously made by several astronomers), he was able to compute the distance to several nearby Cepheids. (He couldn't compute the distance of the Cepheids in the Small Magellanic Cloud this way, because they are so remote they appear to be motionless.)

(Actually, the formula he employed gave inexact distances to any single Cepheid because he couldn't measure the apparent speed of the Cepheid with accuracy; some distances were too long, some too short. He assumed that if he measured the distances to a number of Cepheids, all of which he assumed to be at the same distance from earth, the sum of the inexact long distances would balance out the sum of the inexact short distances, and by taking an average of distances, he could obtain the true distance. What Hertzsprung actually determined was the average formula-distance of a number of Cepheids, all of which he assumed were the same distance from earth. It amounted to determining the distance of a nearby Cepheid of a certain period.)

Knowing the distance of a nearby Cepheid and its apparent magnitude, Hertzsprung calculated its absolute magnitude. A Small Magellanic Cloud Cepheid with the same period has the same absolute magnitude. Knowing the absolute magnitude of that single Cepheid, Hertzsprung worked out the absolute magnitudes of all the Cepheids in the Small Magellanic Cloud. Each Cepheid of a certain period has a certain absolute magnitude.

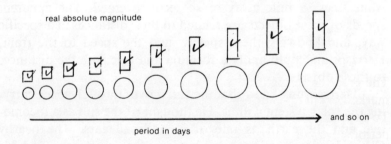

It was easy to estimate the absolute magnitude of Cepheids with periods greater than those of the Cepheids in the Small Magellanic Cloud by extending the line of increasing magnitude in this diagram.

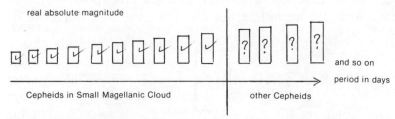

If the height of that line at a certain period was, say, four times the height of that line at a period of 1 day, then the Cepheid of that certain period would have an absolute magnitude four times greater than the absolute magnitude of the Cepheid with a period of 1 day; if the height was ten times higher, it would have a magnitude ten times greater, and so on.

Actually, Hertzsprung drew the line of increasing magnitude as a sloping line on a graph, as shown here:

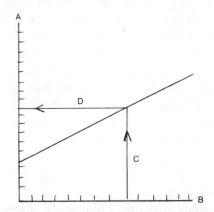

The graph is formed by two perpendicular lines *A* and *B*. *A* is marked off in units of absolute magnitude. *B* in units of period. To determine the absolute magnitude of a Cepheid of a certain period, draw a perpendicular line *C* to the sloping line; then draw another perpendicular line *D* to the line *A*, and read off the absolute magnitude.

Hertzsprung was now close to breaking the code of the blinking stars. With the use of his graph, he could read off the absolute magnitude of any Cepheid in the universe simply by measuring its period. Moreover, he knew that apparent magnitude, absolute magnitude, and distance are related by a formula.

$$\text{apparent magnitude} - \text{absolute magnitude} = \text{distance}$$

The apparent magnitude of a Cepheid could be easily measured by instruments. It was just as easy to measure the period of a Cepheid, from which its absolute magnitude was determined directly. So from two readily obtainable figures—apparent magnitude and period—Hertzsprung could calculate the distance of any Cepheid in the universe. The periods of the Cepheids—the blinks of those stars—were a code helping to spell out stellar distances. Hertzsprung had broken that code.

When Hertzsprung completed his work in 1913, astronomers thought the universe extended no more than 1,000 light-years from earth. How far toward the edge of the universe was the Small Magellanic Cloud? The distance to the Cepheids in it—and, therefore, the distance to the Cloud itself—was, according to Hertzsprung's calculations, an amazing 30,000 light-years. (And Hertzsprung had made an underestimate. Astronomers now know the Small Magellanic Cloud is 150,000 light-years away.)

Cepheids were milestones blinking out immense distances —and astronomers followed them farther and farther into space. What would they find there? The answer to a mystery that had baffled astronomers ever since they first had studied the sky:

What Does the Universe Look Like?

If you stand in an open field and look up at the sky, the six thousand or so stars you can see with your naked eye seem to be painted on the inside surface of an enormous dome. You see that surface in two dimensions, length and width. You can tell where the stars are on that surface,

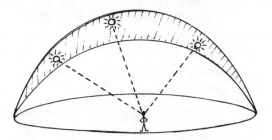

but you can't tell where they are in space.

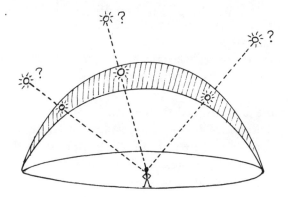

In our view of the stars, the third dimension, depth, is lacking. All stars appear to be at the same height.

Without a three-dimensional view of the sky, astronomers could not know what the universe looked like. Could they convert their two-dimensional view of the sky into a three-dimensional one? There was a way. The missing dimension, depth, was actually the distances to the stars. Astronomers had to find those distances, locate the stars in three dimensions, then build a scaled-down three-dimensional model of the sky with each star in its proper location. Looking at that model would be equivalent to looking at the sky in three dimensions. From the shape of the model, astronomers could tell the shape of the universe.

But before Hertzsprung broke the code of the blinking stars, there was no way to measure the distance to the more remote stars. Now there was. And in 1914 the American astronomer Harlow Shapley set out to get those distances. In 5 years he had obtained enough distances to announce he had found the shape of the universe. But he hadn't. He had found only the shape of a part of the universe—but it was an important part to us, for it is the part in which we live.

The Discovery of Our Galaxy

Shapley worked at Mount Wilson Observatory in California with what was then the world's largest telescope. He trained it on groups of stars called globular clusters, or simply globulars, because they appeared to be globe-shaped. On a photographic plate, a globular looked like a swarm of hundreds of thousands of jeweled bees flying in perfect symmetry. But it wasn't the beauty of the globulars that attracted Shapley to them. It was the high concentration of Cepheids in most of them.

From the periods of the Cepheids and their apparent magnitudes, Shapley calculated the distances to most of the globulars. But how was he to obtain the distances to the globulars that contained no Cepheids? Through a brilliant sequence of observation and reasoning, he invented a new way of determining stellar distances, using the distances obtained by means of the Cepheids as his starting point.

In the globulars whose distances he had already determined, he found very bright stars described by their name red giants. Like all stars in a globular, they were regarded as being at the same distance as the globular. Shapley measured the apparent magnitudes of these red giants and, knowing their distances, calculated their absolute magnitudes. When he examined a long list of the absolute magnitudes he had determined, he observed that all red giants have about the same absolute magnitude.

Shapley reasoned that if he could find a red giant in each of the globulars without Cepheids, he would know the star's absolute magnitude; then by measuring its apparent magnitude, he could determine its distance, which would be the same as that of the globular. With the giant telescope at Mount Wilson, Shapley searched for red giants in those globulars and found them. By 1919 he had determined the distances to about a hundred globulars.

The three-dimensional model of the globulars he made in that year showed them to be arranged in a huge sphere, a superglobular. But where was that superglobular in relation to the rest of the universe? Shapley answered with an inspired guess.

He supposed that just as planets were grouped around the center of the solar system, so the globulars were grouped around the center of the universe. The center of the superglobular was the center of the universe.

Where was the sun in relation to the center of the universe? Since Shapley knew the distance of the sun from the globulars —that is to say, from the surface of the huge sphere—it became a simple problem in spherical geometry to calculate the distance of the sun from the center of that sphere. (Spherical geometry is a branch of mathematics dealing with the measurement of spheres.) His calculations showed that the sun was not the center of the universe, as astronomers of his time thought, but was removed a great distance from the center.

The center of the superglobular is the center of Shapley's universe.

Where were the other stars in the universe? From his knowledge of their distances from the sun, determined by other astronomers, he was able to locate their positions in space relative to the sun. Placing the stars in these positions, he built a three-dimensional model of the universe. Viewed edge on, it looked like a pancake with a bulge at the center.

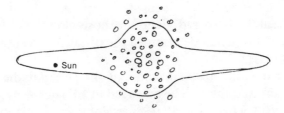

It was as if Shapley had journeyed into space, looked back, and seen the shape of the universe. His discovery was one of the most amazing feats of the human mind.

Most of the stars in the universe were found in or near a softly luminous band of light in the sky called the Milky Way; so modern astronomers named the whole system of stars in the universe after it, using a word derived from the ancient Greek astronomers' name for the Milky Way, *galaxias kylos* (milky circle). That word is galaxy. Shapley's universe was called the galaxy. (Later, to distinguish it from other galaxies, it was called our galaxy.)

Shapley estimated the galaxy to be about 250,000 light-years across and the sun about 50,000 light-years from its center. Between 1919 and 1929 astronomers found a vast body of evidence to support the shape of Shapley's model of the galaxy. But Shapley had been wrong about the galaxy's size. He had not taken into account that dust and gas between stars diminishes apparent magnitude; he had attributed dimness to sheer distance. As a result, he had put the superglobular, and consequently the center of the galaxy, too far away. Corrected figures showed the diameter of the galaxy to be nearly 100,000 light-years and the sun to be about 30,000 light-years from its center.

Two views of the galaxy—Shapley's universe

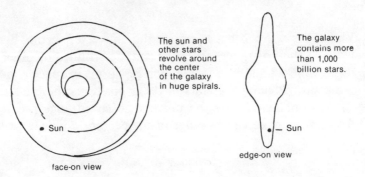

The sun and other stars revolve around the center of the galaxy in huge spirals.

The galaxy contains more than 1,000 billion stars.

face-on view

edge-on view

Shapley thought that outside the galaxy there was nothing but empty space. If Shapley were right, astronomers had reached the edge of the universe, some 20,000 light-years in one direction, some 80,000 light-years in another.

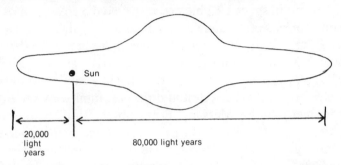

But was Shapley right? The American astronomer Edwin P. Hubble wasn't sure. About 150 years before, the German philosopher and scientist Immanuel Kant had claimed that certain small patches of light in the sky called nebulae (singular nebula) were actually vast clouds of stars distant from the Milky Way. If that were so, astronomers had not yet reached the edge of the universe. There was only one way for Hubble to decide whether Shapley or Kant was right: Look for stars in the nebulae, and if they were found, measure their distances. Hubble swung his telescope, a gigantic new one at Mount Wilson, toward a nebula —and made the first journey to another galaxy.

To the Galaxies

Some nebulae are shaped like disks, others like pinwheels (spiral nebulae), and still others have irregular shapes. The nebula Hubble explored was shaped like a spiderweb. It was named NGC6822. (NGC stands for *New General Catalog,* a list of nebulae and star clusters; 6822 is the number of the listing.) For two years Hubble photographed NGC6822 through his telescope. At the end of that time—it was 1925—he was certain that the spiderweb in the sky contained more than a billion stars. Among them, he identified fifty Cepheids. From measurements of their periods and apparent magnitudes, he calculated the distance of NGC6822. It was an awesome 700,000 light-years—far beyond the limits of what Shapley thought was the universe. Kant had been right; Shapley, wrong.

Far out in space Hubble had discovered a collection of stars as enormous as our galaxy's. It was another galaxy. Were there more galaxies? Hubble turned his telescopic camera to the spiral nebula M33 and found it to be as star-rich as our galaxy or NGC6822. (M33 is the thirty-third entry in a catalog of nebulae, star clusters, and other celestial objects compiled by the French astronomer Charles Messier.) From observations of thirty-five Cepheids in M33, Hubble in 1926 established its distance from the earth at 800,000 light-years. He then focused his telescopic camera on the spiral nebula M31, the Andromeda nebula, and found it was just as heavily populated with stars as the other galaxies. From the sixty-three Cepheids he investigated in the Andromeda nebula, he was able in 1928 to determine its distance. It was 900,000 light-years from earth.

Hubble pressed deeper into space. When his photographs showed no Cepheids, he read off distances from red giants. Within 2 million light-years from earth, he found seventeen galaxies. The farther out he went, the more galaxies he found. As he reached 7 million light-years, he made an astonishing discovery. There he found numerous clusters of galaxies, each cluster containing billions of galaxies and each galaxy containing billions of stars.

But by the end of 1928 Hubble found that his techniques for traveling into space could not carry him farther. The light from galaxies beyond 7 million light-years was too dim to imprint images of stars on the plates of his telescopic camera; and without observations of Cepheids or red giants, distances could not be determined. Then 1 year later Hubble made another discovery. It would enable him to journey through distances that would dwarf the 7 million light-years through which he had already traveled. The clue to that discovery came from a man-made rainbow.

The Clue of the Speeding Galaxies

Light travels in a wave motion.

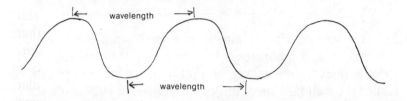

The distance between peaks or troughs of the wave is the wavelength. Light is composed of separate wavelengths, which our eye recognizes as different colors. The spectroscope is an instrument that spreads out a beam of light into an array of different colors, a man-made rainbow, known as the spectrum (plural spectra). Violet, the shortest wavelength of visible light, is at one end of the spectrum; and red, the longest wavelength, is at the other.

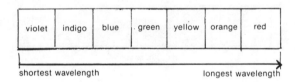

The short-wave end of the spectrum is called the blue; the long-wave end, the red. Scientists mark off the spectrum in units of increasing wavelength.

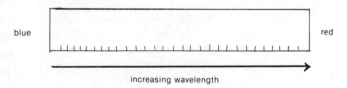

The spectrum of light from a galaxy shows a series of lines at specific wavelengths. Each of these spectral lines, or a pattern of them, signals the presence of a certain element in the galaxy.

Spectral lines are also obtained when small samples of elements are burned in a laboratory. By comparing the position of the spectral lines from a galaxy with those of the laboratory sample (called the standard spectrum), astronomers in 1928 could tell in which direction a galaxy was traveling and how fast.

If a galaxy were moving toward the earth, its spectral lines would shift toward the blue.

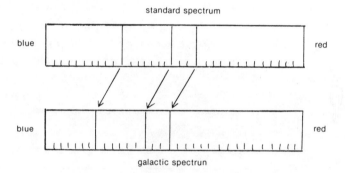

If the galaxy were moving away from the earth, its spectral lines would shift toward the red.

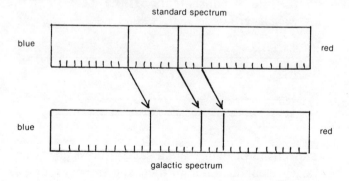

By measuring the amount of shift in the wavelength, astronomers could determine the velocity of the galaxy.

What happens is this: When a light source and an observer are stationary, the wavelength of light is unaffected.

But when the light source is moving away, the light waves are stretched out as if the light source were pulling them apart.

Each wavelength becomes longer. So the wavelength of violet lengthens into indigo; the wavelength of indigo lengthens into blue, and so on; and the entire spectrum shifts toward the red.

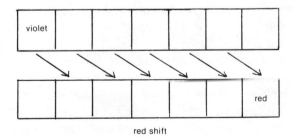

red shift

When a light source is approaching an observer, the light waves are compressed as if the light source were squeezing them in the direction of the observer.

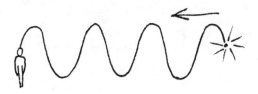

Each wavelength becomes shorter, and the entire spectrum shifts toward the blue.

Just how much the wavelengths are stretched or squeezed—that is to say, just how much the spectrum is shifted toward the red or blue—depends on the speed of the light source. The greater the speed, the greater the shift. By measuring the amount of shift, the speed of the light source can be determined. (The relation of wavelength to motion had been discovered by the Austrian physicist Christian Johann Doppler in 1842 and is known as the Doppler effect.)

Hubble had been interested in the speed of nebulae since 1914, when he heard the American astronomer Vesto Melvin Slipher at an American Astronomical Society meeting report an extraordinary discovery. Slipher had analyzed the spectra of about twenty nebulae and had found to his surprise that every spectrum shifted to the red. That meant all twenty nebulae were moving away from us. From the amount of the red shift, he had calculated that the speeds of these nebulae ranged up to several million miles per hour, the greatest speeds observed in the heavens up to that time.

Hubble knew that the nebulae Slipher had investigated were actually galaxies. Were other galaxies moving away from us—astronomers preferred to say receding—at very high speeds? Or was Slipher's surprising find a freak discovery? From 1928 to 1929 Hubble and his associate Milton Humason photographed the spectra of many galaxies (a spectroscope was connected to the giant telescope at Mount Wilson) and found that Slipher's discovery was not a freak. All the spectra they examined were shifted to the red; and the large amount of the shifts indicated that some galaxies were receding at speeds up to 100 million miles per hour.

Slipher had concluded that the more remote the nebula (galaxy), the faster its speed. This clue of the speeding galaxies gave Hubble the idea that the distance of a galaxy might be determined from its speed. Slipher had only been able to estimate comparative distances based on the different dimnesses of the galaxies he had observed, but Hubble had obtained actual distances of those galaxies and many others as well as their speeds. He set out to check Slipher's conclusion about the relationship between the speeds and distances of galaxies.

When Hubble listed the speeds of galaxies in one column and their corresponding distances in another, he saw at once that Slipher's conclusion was correct. Moreover, distance varied with speed according to a specific relationship. A galaxy speeding away twice as fast as another was twice as distant as the other; a galaxy speeding away three times as fast was three times as distant, and so on. This relationship is known as Hubble's law.

With this new law Hubble could calculate the distances of galaxies. For example, a galaxy with a speed two times greater than that of a galaxy 7 million light-years distant would be 14 million light-years away; a galaxy with a speed three times greater would be 21 million light-years away, and so on. Hubble expressed his law in a simple mathematical formula that enabled him to determine the distance of a galaxy from its speed. Then, since the speed of a galaxy is measured by the amount of its red shift, he derived another formula that gave him the distance of a galaxy directly from its red shift.

It was no longer necessary to find Cepheids or red giants in a galaxy in order to establish its distance. The red shifts had become the new milestones of the universe.

Hubble and Humason photographed the spectra of dim galaxies and began to probe into the immense expanse of space.

By the early 1930s they had charted galaxies hundreds of millions of light-years away. Then for nearly 2 decades they plunged deeper and deeper into the remote regions of the universe. By 1950 they had identified galaxies at a distance of 2 billion light-years. But no matter how carefully they searched the sky with their telescopes, they could find no galaxy farther away. Had the edge of the universe been reached?

Some astronomers speculated that more distant galaxies did exist but that they were so far away they were invisible. If that were so, the edge of the universe would be lost in the darkness of outer space. How were astronomers to penetrate that darkness?

Seeing the Invisible Universe

By 1950 astronomers had long been accustomed to using two instruments, the telescope and the photographic plate, to make invisible objects in the sky visible. These instruments were much like the human eye in principle but far superior in performance.

The human eye sees when a definite amount of light activates a screen called the retina at the back of the eye. When not enough light is received from an object, the retina is not activated and the eye cannot see the object. To obtain the necessary amount of light, the eye is equipped with a device that collects light. This is a flexible opening in the front of the eye called the pupil.

In a darkened room the eye begins to see objects only after a few seconds because the pupil must widen to collect more light. The larger the pupil, the more light it collects, and the fainter the objects that can be seen.

To the human eye, the night sky is very much like a darkened room. At first the eye receives only enough light to see the brighter objects in the heavens. Then as the pupil widens, the eye is able to pick out fainter and fainter ones. But there is a limit to the size of the human pupil. The reason the naked eye can make out only about six thousand stars is that the pupil cannot grow large enough to collect sufficient light from any of the other billions and billions of stars to activate the retina.

To see those stars and other invisible objects in the sky, astronomers make use of an artificial pupil immensely larger than the human pupil—the telescope. Contrary to common belief, a telescope is rarely used only to magnify an image; it is used primarily to collect light. Astronomers think of a telescope as a bucket into which light is poured; and in scientific slang it is known as a light bucket. The larger the light bucket, the more light it collects; and the more light it collects, the fainter the objects astronomers can see in the universe that is invisible to the naked eye.

A giant light bucket collects light by means of a huge mirror. The light is reflected with extreme accuracy and focused on a magnifying system. The size of a giant light bucket is designated by the size of its mirror. A light bucket with a mirror 60 inches in diameter is known as a 60-inch telescope; a light bucket with a mirror 100 inches in diameter is known as a 100-inch telescope, and so on. Shapley discovered our galaxy with a 60-inch telescope, and journeyed 80,000 light-years from earth. Hubble discovered the galaxies with a 100-inch telescope and plunged 2 billion light-years into space. The largest telescope ever built—the telescope that would take astronomers to the edge of the universe—is the 200-inch at Mount Palomar Observatory, California. It weighs more than 100 tons and collects 500,000 times more light than the human pupil.

The light collected by the giant telescopes is not normally viewed directly by the human retina. Instead, astronomers use an artificial retina, the photographic plate. It has a quality the human retina does not possess. It can see invisible objects by prolonged staring (exposure), whereas the human retina cannot.

The human retina is affected only by the light it receives from an object in the first .1 second of viewing. After that, no matter how much light from the object strikes the eye as a result of prolonged staring, the retina will not become activated. It has no capacity to add up light. But the photographic plate does. As it stares at an object (as it's exposed to an object), the light it receives grows until there is a sufficient amount to activate the chemicals on it and an image of the object appears. What is invisible to the naked eye becomes visible on the photographic plate. An astronomer viewing a region of the night sky through the eyepiece of his telescope might see only a spot of darkness, but a photographic plate staring for some time through the telescope at that spot of darkness might see the faint image of a galaxy in the depths of space.

With the giant telescope and the photographic plate, the astronomers of the 1950s had the instruments for searching the invisible universe for galaxies farther out than 2 billion light-years. But where in the sky should they point their telescopic cameras to look for them? The sky is enormous, and to probe every black spot of it would take teams of researchers innumerable lifetimes. Yet by 1950 astronomers had found a way to know exactly where to spot a galaxy in the invisible universe. The clue to that way had come in 1933, when Karl Jansky, an American engineer, tracked down the source of a strange hissing sound in his equipment for receiving radio waves.

The Clue of the Radio Waves from Space

More than 50 years before Jansky stumbled on the clue, the Scottish physicist James Clark Maxwell had made a series of breakthrough discoveries concerning the electromagnetic force. The force produced energy in the form of rays—electromagnetic radiation—that traveled through space as waves, just as light waves did. As a matter of fact, light rays were one type of electromagnetic radiation. The other types were invisible. At the time Maxwell made his discovery, two kinds of this invisible radiation were known: the ultraviolet and the infrared,

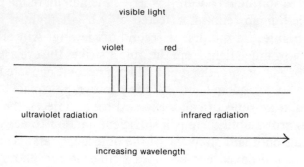

the former beyond the violet end of the spectrum (*ultra* is Latin for "beyond") and the latter coming below the red end of the spectrum (*infra* is Latin for "below"). The wavelengths of infrared radiation were longer than those of the visible part of the spectrum; the wavelengths of ultraviolet radiation were shorter.

Maxwell predicted that electromagnetic radiation of other wavelengths would be found. In 1889, a decade after Maxwell made that prediction, the first such radiation was produced in the laboratory of the German physicist Heinrich Rudolph Hertz. At that time telegraphy was the major means of communicating at a distance. The operator of one telegraph would tap out a message in a code of electrical impulses that would be carried over a wire to another telegraph that would convert these impulses into clicking sounds. Combinations of long and short clicks stood for the letters of the alphabet. In 1896 the Italian inventor Guglielmo Marconi found that the radiation Hertz had discovered could produce the clicking sounds in a telegraph without the use of wires.

This was wireless telegraphy, and the radiation that made it possible was called radiotelegraphic waves (radi- coming from the first part of the word radiation). The name was soon shortened to radio waves. Compared with the wavelengths of light waves, those of radio waves were enormous.

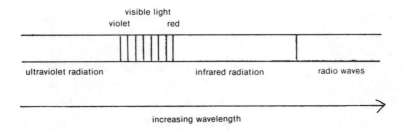

(Later, physicists found several kinds of radiation on the other side of the spectrum. The complete spectrum of radiation produced by the electromagnetic force is called the electromagnetic spectrum, and it stretches from the very short waves of cosmic rays to extremely long radio waves. Visible light occupies only about less than 2 percent of the electromagnetic spectrum. All radiation is electromagnetic, so 98 percent of the radiation in the universe is invisible to us.)

In 1921 radio waves were first used to reproduce the human voice and other sounds in a telephone receiver, and that was the start of radio as we know it today. A little more than a decade later the Bell Telephone Laboratory at Holmdel, New Jersey, was beginning to experiment with shortwave radio (radio waves shorter than most other radio waves) as a replacement for cable in the transmission of telephone messages. And Karl Jansky, an employee of Bell, was given an interesting assignment.

The researchers at Bell were concerned about possible interference with shortwave radio messages due to thunderstorms. It was Jansky's job to tune in a radio receiver on thunderstorms and listen to the radio noises they produced. To do it, he had to have a radio receiver that would pick up radio waves wherever thunderstorms appeared in any part of the sky. No such receiver existed, so he improvised one.

He took the rear of a Model-T Ford, and turned it sideways so that one wheel faced upward and was free to rotate. To this wheel he attached an antenna that consisted of a number of steel rods shaped like squared-off inverted U's. By swinging the wheel around full circle, he could aim his antenna at all parts of the sky. The antenna fed into his radio-wave receiver, which was not a radio in the ordinary sense but a set of exposed telephone cables. These were connected to a moving-paper recorder and an amplifier.

When thunderstorms appeared, the amplifier poured out a stream of explosive pops, sharp snapping sounds, and the familiar crackle of static—the radio noises of thunderstorms. During certain parts of the day he also heard a peculiar hissing sound even when there was no thunderstorm. Where did this sound come from?

Jansky carried on his observations for 2 years. Each day the hissing sound came from just one source, and that source was always in the same position on the horizon. Could that source be in the atmosphere? Not likely, Jansky thought. The atmosphere is a swirling, turbulent torrent of gas; nothing is ever in the same position in it. But if the source was not in the atmosphere, it must be outside of it, somewhere in space. Where?

Painstakingly, Jansky narrowed down the source to one region in the sky. In 1933 before a meeting of the American Section of the International Radio Union, Jansky announced that the radio waves producing the hissing sound in his radio were coming to earth from the core of our galaxy.

Jansky's improvised setup on the rear wheel of a Model-T Ford was the first radio telescope, an instrument for detecting radio sources in the heavens. The radio telescope operates basically very much like the older type of telescope, the optical telescope. Just as the optical telescope is thought of as a bucket for collecting light waves, so the radio telescope is thought of as a bucket for collecting radio waves. In most radio telescopes a metal dish acts as a mirror, focusing the radio waves it collects onto an antenna that is connected to a sensitive radio receiver. Because radio waves are enormously longer than light waves, radio telescopes are ordinarily huge compared with their optical counterparts. Optical telescope mirrors are measured in inches, but the "mirrors," the dishes, of radio telescopes are measured in feet. (In 1979 the University of California had begun planning what will be the world's largest optical telescope. Its mirror will be 400 inches in diameter. The dish of the world's largest radio telescope at Arecibo, Puerto Rico, is 1,000 feet wide.)

As the 1950s began, mammoth radio telescopes were starting to swing into action in the United States, Australia, the Soviet Union, and Britain. They could sweep the sky far more rapidly than optical telescopes; and sources could be detected without the long waits astronomers had to endure while photographic plates were adding up light. With the radio telescope, astronomers had found a way to know where to look for an object in the invisible sky. The radio telescope would find the position of a radio source; then the camera of the optical telescope would be trained on that position.

In 1951 astronomers at Cambridge University, England,

pinpointed the position of the radio source Cygnus A. The American astronomer Walter Baade focused the long-exposure camera of the 200-inch telescope on it and discovered the first radio galaxy. The measurement of its red shift showed it to be 3 billion light-years away—a billion light-years more distant than the most remote galaxies discovered by Hubble and Humason.

In 1959, with radio telescopes doing the spotting for him, Humason, using the 200-inch telescope, discovered a galaxy at a distance of 4.9 billion light-years. In 1960 radio telescopes clued the American astronomer Rudolph Minkowski to turn the 200-inch telescope to a black spot in the sky where he photographed a galaxy with the largest red shift yet found, indicating a velocity of half the speed of light and a distance of 7 billion light-years. And radio astronomers were locating more and more radio sources. Would these sources lead astronomers to the edge of the universe?

Four radio sources would provide a strange puzzle. The Holland-born American astronomer Maarten Schmidt would solve it, and astronomers would start on their last lap to the edge of the universe.

The Puzzle of the Mysterious Spectra

In searching for distant radio sources, astronomers were guided by this rule of thumb: If the source is limited to a small patch of sky—that is to say, if the source is compact—it is likely to be extremely far away. The same year that Minkowski penetrated 7 billion light-years into space, the American astronomer Allan Sandage used the 200-inch telescope to photograph a compact radio source 3C48 (number 48 in the third catalog of radio sources compiled by Cambridge University). The developed plate revealed a small blue point of light that seemed to be a star.

A single star at the great distance indicated by the compactness of the source had never been found before; only galaxies were known to exist in the far distant reaches of space. The fact that a star existed there was strange enough, but the spectrum of the star was even stranger. Every star was known to have a definite pattern of spectral lines indicating the presence of certain elements. But the pattern of lines in 3C48 was like nothing Sandage had ever seen in a star before. "It was the weirdest spectrum I ever saw," Sandage said. "I tried very hard to identify the spectral lines of 3C48 but I was baffled."

In the next 2 years three other radio sources that appeared to be stars were found. Like 3C48, they exhibited spectral lines that were unidentifiable. Did that mean distant stars were made up of elements unlike any elsewhere in the universe? It was a speculation that flew in the face of everything scientists had learned about the makeup of matter. Maarten Schmidt looked for another explanation. Spectroscopy (the study of spectra) was his specialty, and "Any puzzle in spectroscopy," he said, "I think of as a challenge."

But he was unable to beat the challenge of the four unidentifiable spectra. Then in 1962, using the 200-inch telescope, Schmidt photographed the spectrum of a fifth radio source that looked like a star. The photographs he took, Schmidt wrote later, "finally gave the clue to the mysterious spectra of the 'radio stars' as they were called at that time."

On the photographs Schmidt identified a well-known pattern of spectral lines produced by the presence of hydrogen. These lines appeared in the spectra of nearby stars in this position.

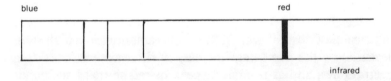

But the spectra of the fifth radio star, 3C273, showed these lines far to the red.

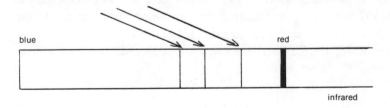

Had the hydrogen lines red-shifted? It seemed incredible. Up to that time—it was 1963—"it had been faint galaxies exclusively which showed red shifts of this size," Schmidt wrote later, "but these new objects [the radio stars] looked like bright stars!" If the hydrogen lines had actually shifted to the red, Schmidt thought, the spectral lines at greater wavelengths must have shifted into the infrared.

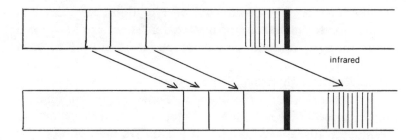

Though infrared radiation cannot be seen by the human eye, it can be seen by specially sensitized film. When Schmidt photographed the infrared spectrum of 3C273, he found that spectral lines from the visible spectrum had been shifted into the infrared exactly as he thought they would be. Here was an object at an enormous distance from earth speeding away like a galaxy—and it looked like a star. "This development was astounding," Schmidt wrote, "and it took a few days to recover from the shock."

Just how distant was 3C273? The amount of red shift is measured by the percentage increase in the wavelengths of the spectral lines; and astronomers speak of red shifts of such and such a percent. The red shift of 3C273 was 16 percent, about the same as that of galaxies about 1 billion light-years from earth.

The red shifts of the first four radio stars were even greater —so great that most of the visible spectral lines had been shifted into the infrared, accounting for the weird visible spectra that had been observed. The most distant of the first four radio stars was 3C48, the first of them to be discovered. It was 4 billion light-years from earth.

Schmidt decided to find out more about these unusual stars. He began by studying their brightness. Since they were so far away, for them to appear as even a faint point of light on a telescopic photograph, they would have to be as brilliant as a hundred average-size galaxies. Were they—incredible as it seemed—as big as a hundred average galaxies?

Even though the radio stars appeared as little more pinpoints on his photographic plates, Schmidt could estimate their size. The light from radio stars varies in brightness over definite periods of time; and astronomers knew that the shorter these periods, the smaller the star. Schmidt found the periods of varying brightness of the radio stars were extremely short. From that he concluded that far from being supergalaxies, radio stars were "remarkably small with typical diameters of perhaps as little as a few light-years . . . . They are several thousand times smaller than a typical galaxy." But they were several million times larger than a typical star.

Too large to be stars, too small to be galaxies, and yet as bright as a hundred average-size galaxies—what were these strange distant objects? "Mysterious objects masquerading as stars" was the only answer Schmidt could give. Astronomers called them quasars. (The word is a contraction of quasi-stellar objects. *Quasi* is Latin for "resembling," and quasi-stellar means "resembling a star.")

By solving the puzzle of the mysterious spectra, Schmidt had discovered a new kind of celestial object as distant as galaxies. Astronomers began to search for more quasars, and as they found them, the edge of the universe grew closer and closer.

The Final Lap

In 1964 Schmidt found quasars at a distance of 9 billion light-years—2 billion light-years farther out than the most remote galaxies. In the next several years other astronomers discovered quasars at distances of between 12 to 14 billion light-years. By the early 1970s red shifts had been measured for some two hundred quasars. The most remote, receding at 70 percent of the speed of light, were 17 billion light-years from earth.

But there were only a few of these extremely remote quasars. It seemed as if the quasars were thinning out with distance. As time passed, neither radio nor optical telescopes could detect quasars farther away than 17 billion light-years. Beyond that limit there appeared to be nothing. And the *New York Times,* reporting on the most distant quasars, headlined the story with: "Men Report Seeing Edge of the Universe."

The date was April 8, 1973.

But was the edge of the universe where astronomers saw it? The light traveling from a quasar 17 billion light-years away took 17 billion years to reach the earth. What astronomers were observing was the edge of the universe as it appeared 17 billion years ago. Was it still at the same distance from earth? Or, during the vast stretch of time, had it moved away from us like the expanding galaxies? Scientists searched for the answers, and the answers to other cosmic riddles, as they traveled through time, from its beginning in the remote past to its end in the distant future.

5

The Journey from the Beginning to the End of Time

The announcement of the completion of the first machine designed to travel through time was announced in 1895 by its inventor at a dinner party at his home in London. The guests, a small group of distinguished men in various fields of activity, viewed the announcement with polite skepticism. The inventor, who is known to history as the Time Traveler, led the guests to his laboratory located in his home and showed them a delicately made mechanism constructed of glittering metal and some transparent substance. It was equipped with a saddle much like a bicycle seat, in front of which was an array of dials and levers.

When one guest asked his host whether he was serious or whether he was playing another joke on them like the ghost he showed them the previous Christmas, the Time Traveler replied that he intended to explore time in that machine and that he had never been more serious in his life. Not one of his guests believed him.

A little more than a month later, the guests again met for dinner at the Time Traveler's home, but their host was absent. He had left a note saying that he was unavoidably delayed, that they should go on with dinner at seven if he wasn't back, and that he would explain later in the evening. Toward the end of dinner, one of the guests, an editor of a publishing house, was going on at length about the absurdity of time travel and how the Time Machine was a hoax when the door leading from the Time Traveler's laboratory opened quietly and the Time Traveler stood on the threshold of the dining room.

He was in a terrible condition. He was dusty and dirty. His face was ghastly pale, and his expression was distressed and pained as if he had undergone serious trouble. There was a half-healed cut on his face, and his hair seemed to have turned grayer. He limped into the room and without saying a word sat down at the table and motioned for food.

When he had refreshed himself with champagne and roast beef, he told his guests that he had lived the last 8 days as no human being had ever lived before. What he was about to tell them, he said sincerely, might sound like lying, but it was true—every word of it. Then in semidarkness, for most of the candles in the room had not been lit, the Time Traveler told the following story.

I drew a breath, set my teeth (the Time Traveler said), pressed the starting lever of the Time Machine and went off with a thud. The night came on suddenly, and in another moment tomorrow dawned; then night again, day again, faster and faster until the change of light and darkness became a twinkling in the eye. The little hands on the dial that registered my speed in years twirled around faster and faster. When I pulled on the stopping lever and read the dial, I found that I had traveled to the year 8,003,701 A.D.

I found myself on the hillside where my home in London had been situated so many millions of years ago. Ahead of me was a colossal figure carved in white marble, shaped like a winged sphinx, a creature with a lion's body and the head and bust of a woman. The wings were spread and seemed to hover over a bronze pedestal on which the sphinx rested. In the distance I saw the tall columns of huge splendid buildings. Then I heard voices of men approaching me.

The first man of the distant future I laid eyes on was a beautiful and graceful creature about 4 feet high and unbelievably frail. The others, eight or ten of them, were just as exquisite. They were dressed simply in tunics and sandals, for the air was very warm.

Laughing merrily, they led me to their home, a vast palace-like building. In a great hall, a large number of marble tables were heaped with fruit, for these delicate people of the future were strict vegetarians. About a hundred of these people of both sexes were in the hall, and the air was filled with their joyous voices. It seemed to me that they were a playful and happy people without a care in the world. They were, as I soon learned, descendants of the very rich. Apparently others cared for their every need, and their lives consisted only of fun and frolic.

Yet something was wrong. The palace in which they lived, magnificently designed as it was, seemed to be wearing away with age. The stained-glass windows were broken, the marble tables were fractured, and the lovely curtains were torn and thick with dust. The people themselves seemed without mental curiosity, even stupid or retarded. They could not read or write. And above all, they had a strange, deep fear of the dark, and of the marble sphinx in front of which I had left my Time Machine.

My Time Machine! Was it still there? I hurriedly retraced my steps. There was the sphinx, but my machine was nowhere to be found. I was gripped with a fear that I would be stranded in this strange world of the future, never again to see my home and friends. I searched everywhere, but not a trace of my machine could I find. The beautiful people—their name, by the by, is the Eloi—seemed to know nothing about its disappearance.

I had made a friend of a beautiful young female Eloi—really no more than a child intellectually, but a sweet and affectionate creature. She had been carried away by a swift current while bathing in the river and, while the other Eloi looked on indifferently as she was being swept to her death, I had plunged in and brought the poor bedraggled darling back to the safety of the shore. Weena is her name, and after her rescue she became my inseparable companion. As a gesture of her gratitude, she used the pocket of my jacket as a flowerpot—for she had never seen a jacket before and thought the pocket could only be a flowerpot—and filled it with beautiful white flowers unknown in our time.

With Weena I searched the countryside for my Time Machine. As I did, I came across a number of wells, the sight of which made Weena turn pale. Looking into them I could only see an immense darkness, but I could hear the thud and murmur of machinery. So that was where the Eloi received their clothes and other necessities, I thought—from an underground city of workers. Had the workers stolen my Time Machine? Was there an entranceway to the underground city through a door in the pedestal under the sphinx, and had the workers dragged my machine behind that door?

One hot morning to escape the glare of the sun, Weena and I took refuge in a cool dark passageway of a ruined building. A queer ape-like creature suddenly brushed past us. It was covered with dull white fur and had the overlarge lidless eyes of a creature who lives in the blackness of caves. As it loped out into the daylight, shielding its eyes, we followed. It made for one of the wells and disappeared into it. Looking down the shaft, I saw this foul animal, like a human spider, clambering down the ladder. That monster, I realized, was a worker, an inhabitant of the lightless underground city. It appears that in the millions of years from our time, the human race had evolved into two distinct kinds of animals, the graceful children of the upper world, and that horrible Thing who lived in the depths.

That Things like him had stolen my Time Machine, I had no doubt; and despite Weena's pleadings, and armed only with a package of pocket matches, I descended into the well. When I reached bottom, I saw with the aid of my lit matches—for I was otherwise in utter blackness—that I was in a world of machinery run by these Things. It was clear to me that beginning with the construction of the subways in our big cities, the ancestors of the Eloi had banished their ugly means of transportation, their factories, and finally their workers to the underground where they would be forever out of sight. But it seemed to me strange that the Things the workers had become should continue to serve the childlike Eloi above. Then I saw the meat.

The Things, which are called Morlocks, were eating it; and there was no doubt what it was. The horrible truth about the Morlocks and the Eloi dawned on me. The Morlocks were tending the Eloi like cattle. When night fell, the Morlocks would emerge from their wells and drag the Eloi below to be slaughtered. This was the reason the Eloi lived in dread of the dark and of the wells.

As long as my matches lasted I was safe from the Morlocks for, accustomed to the darkness, light blinded them. My last match flickered out, and they attacked me. I fled and scuttled up a well ladder to safety. But I knew that when darkness fell, they would come after me in great numbers, for I had had to fight my way out and I had injured many of them.

When they streamed out of their wells that night, I was ready for them. I had found some matches in my pockets, and I had set the foliage ablaze. Blinded by the fierce light, many of the Morlocks ran straight into the inferno and perished; others, somehow, stumbled back to their wells and retreated into them. I was safe from the Morlocks. But how could I retrieve my Time Machine from them?

With Weena clasping my hand, I returned to the lawn in front of the sphinx. I stood there racking my brains for some means to force my way through the bronze pedestal when its doors swung open revealing my machine. By the moonlight shafting through the entrance-way, I could see it had been unharmed. The Morlocks, masters of machinery, had even cleaned and polished it; and there it stood, ready to go.

I rushed through the entrance-way, Weena following at my heels, and the doors clanged shut behind me. The wily Morlocks, knowing that my defenses were too strong for them in the upper world, had trapped me in their underground, using my machine as bait. At once, they snatched Weena and carried her away while her screams rang in my ears. It would have been futile to go after her for I had only one match left. Holding the Morlocks at bay with the last gasp of matchlight, I leaped into the saddle of the Time Machine and slammed the starting lever.

And here I am, gentlemen (the Time Traveler concluded). And now, I must go back to my machine.

On the following day, one of the guests returned and made his way directly to the laboratory. As he opened the door, he heard a click and a thud, and a gust of air whirled him around. He seemed to see the Time Machine as a faint blur and the Time Traveler as a transparent figure seated on the saddle. Then, as the guest rubbed his eyes, the vision vanished.

Had the Time Traveler gone forward in time to the land of the Eloi and the Morlocks? Had he returned there to attempt to rescue Weena, knowing that the other Eloi would be utterly indifferent to her fate? Or, the guest continued to wonder, had the Time Machine ever existed at all?

There was one small bit of evidence to indicate that it had. During the course of his tale, the Time Traveler had taken two strange beautiful white flowers out of his pocket—the flowers he had claimed Weena had placed there as in a flowerpot after he had saved her life. Years later, when all hope for the Time Traveler had passed, the guest looked at these flowers—shriveled now, and brown and flat and brittle—with a sense of comfort. They were proof to him that even if human beings were to degenerate into fools and beasts, tenderness and gratitude would still live on in the heart of man. And knowing that, he knew there would always be hope for mankind. THE END.

H. G. (Herbert George) Wells, the eminent English writer, published *The Time Machine* in 1895. In the opening section of this first and greatest of all stories of time travel Wells introduced the idea of time as a dimension like length, width, and depth—a fourth dimension—an idea that did not receive wide scientific acceptance until Einstein two decades later made it an essential part of his picture of how the universe works.

Since the publication of *The Time Machine*, travel in the fourth dimension—which really means travel in time—has been a favorite theme of science fiction. But no fictional time traveler has ever had a goal as ambitious as that of the real-life scientists who traveled (within their minds, of course) through time. That goal was to journey from the beginning to the end of time. When that goal was achieved in the 1970s, science had taken its most fantastic journey.

During the course of that journey, scientists witnessed the birth of the universe and its death. They saw the marvels of creation—the formation of quasars, stars, galaxies, planets, and the first stirrings of life. They watched the solar system take shape and the sun age and die. They were there when stars exploded and shrank to dwarfs of their former selves. They looked on as the most astounding objects in the universe came into being: black holes, bottomless pits of force that swallowed everything that approached them, even light. They traveled 18 billion years into the past and 67 billion years into the future.

But before the journey could start, scientists had to know whether or not time had a beginning. The clue to the answer came from a discovery so strange and surprising that the great scientist who made it refused to believe it.

The Clue of the Expanding Universe

That scientist was the German-born physicist Albert Einstein. In 1915 Einstein worked out his general theory of relativity. It was an explanation of the movement of celestial objects, based on radically new concepts of space and time.

A scientific theory, such as Einstein's, must not only agree with observed facts, it must also be able to predict facts that have not yet been observed. Einstein's theory predicted that astronomers should be able to observe a universe in which galaxies were flying apart from one another—an expanding universe. But astronomers searching the heavens with their telescopes could only observe a universe that was at rest and unchanging—a static universe. In the face of this evidence Einstein came to regard his discovery as an incredible mistake, and he changed his theory to conform to the concept of a static universe.

But Einstein's idea intrigued the Dutch astronomer and mathematician Willem de Sitter. Einstein had expressed his theory in mathematical symbols. The ways the symbols were arranged formed a code of sorts—a series of mathematical equations—that revealed some information on sight but contained more information that was hidden even from Einstein, who had invented the code. By rearranging the symbols according to the rules of mathematics—that is to say, by solving the equations—de Sitter was able to read some of the hidden information.

He found that, even with the change Einstein had made in his equations, the universe could expand, but only if it were composed of nothing but empty space. Since the universe was composed of matter and energy as well as space, de Sitter regarded his finding as having no relation to reality—a mathematical fiction. But he hoped it would stimulate the thinking of other scientists.

One such scientist was the English astronomer and mathematician Arthur Stanley Eddington. De Sitter had described his vacant expanding universe in the form of equations, and Eddington wondered what would happen if he introduced symbols that stood for galaxies into those equations. When he solved the equations containing those symbols, he discovered that all the galaxies in the universe should be flying apart from one another. But Eddington's equations were based on de Sitter's equations, which in turn were based on Einstein's quotations, and Einstein interpreted his equations as describing a static, not an expanding universe. Who was reading Einstein's mathematical code correctly—Einstein or Eddington?

In 1922 the Russian mathematician Alexander Friedman began a careful reexamination of Einstein's work. He was startled to find that Einstein had made a schoolboy error. Einstein had divided by a symbol that could stand for zero. Dividing by zero is forbidden by the rules of mathematics. When Friedman corrected Einstein's mathematics, then translated the mathematical symbols into the real forces and things they represented, he found that it was Eddington and not Einstein who had read Einstein's mathematical code correctly. According to Friedman's reading of that code, the universe was expanding.

But in 1922 most astronomers held firm to the idea of a static universe. They were unaware that in 1914 Vesto Melvin Slipher (whose findings had been the basis for Hubble's law) had discovered about twenty nebulae moving away from earth at very high speeds. Nor did they know that between 1914 and 1922 Slipher had extended his list of receding nebulae to thirty-six, providing a strong hint that the universe was growing larger. Without knowledge of Slipher's discoveries, and with no other tangible evidence to support the conclusions Eddington and Friedman had drawn from Einstein's general theory of relativity, astronomers dismissed the idea of an expanding universe, as de Sitter had, as a mathematical fiction.

But Eddington knew of Slipher's work and clung to the

idea. In 1927, to arouse scientific interest in the concept of the expanding universe, Eddington published Slipher's table of nebulae red shifts, solid facts that proved some nebulae were speeding apart. Slipher's table found its way to the desk of the Belgian mathematician Abbé George Lemaître. Lemaître, not knowing of Friedman's work, had analyzed Einstein's mathematics and had come to the same conclusion as Friedman. The universe *was* expanding, and Lemaître suggested that confirmation would be found by more studies such as Slipher's.

In the United States Edwin Hubble was making just such studies. In 1929 he published his milestone papers providing overwhelming evidence that nebulae—now known to be galaxies—were receding from earth and from one another at tremendous speeds. (Papers are scientific reports published in magazines—scientific journals—read by scientists. Published papers are the basic means by which scientists share their discoveries with one another.)

Hubble's observations clearly confirmed the prediction of an expanding universe made by Einstein's general theory of relativity, a prediction that one expert has called "the greatest . . . in the history of science." But like Einstein, Hubble was reluctant to accept the idea of a universe that was still growing. "The notion," he said, "is rather startling."

But no one could dispute Hubble's findings, and by the early 1930s the scientific community had accepted the expanding universe as an established fact. People who were not scientists found the idea of galaxies moving away from one another hard to grasp, and some scientists explained it to them this way: Just as raisins in a cake dough move away from one another as the dough rises in a baking pan,

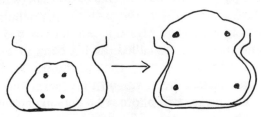

The arrow represents time.

so do the galaxies move away from one another as the universe rises in space.

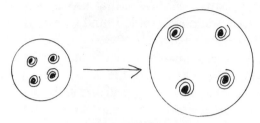

The arrow represents time.

But the heat of an oven causes the dough to rise in a pan; what kind of energy caused the galaxies to rise in space? The energy of an explosion, Lemaître speculated. In 1932 he proposed that at one time all the galaxies in the universe were packed tightly together in what he called a primordial atom perhaps 200 million miles in diameter. (*Primordial* means "first created.") This atom, for some unknown reason, exploded, sending the galaxies flying apart and creating the expanding universe.

In 1948 the Russian-born American physicist George Gamov applied his knowledge of atomic physics to Lemaître's speculation. Gamov conceived of a primordial atom less than a few miles in diameter, composed not of galaxies but of a dense glob of neutrons. The glob exploded—Gamov couldn't tell why—and produced fundamental particles of matter and energy that were hurled into space.

These particles, Gamov thought, then acted on one another to produce elements, which gradually combined to form galaxies. The original force of the explosion was carried from the fundamental particles to the elements to the galaxies, driving the galaxies apart. Gamov called the explosion that had created the universe the big bang; and his explanation of events before and after the universe began is called the big bang theory.

According to the big bang theory, the primordial neutron glob was changeless. There was no time then, because time is change. When you see two photos of a person, one at the age of 10 and the other at the age of 30, you might say, "There's been quite a change. One of these photos was taken after the other. Time passes." But if you were able to photograph the primordial neutron glob, all the photographs would look the same no matter how many you took. Examining them, you might say, "I can't tell which photo was taken after the other. They seem to have been taken at the same time." When everything is always at the same time, there is no time.

But, the big bang theory proposes, if you were to take photos after the big bang, one photo would show particles farther apart than in another photo. Things had begun to change, and time had begun to pass. With the big bang the universe was born, and so was time. The clue of the expanding universe had brought Gamov to the conclusion that time did have a beginning. But was Gamov right—had there been a big bang?

Fred Hoyle, Herman Bondi, and Thomas Gold didn't think so. The same year Gamov announced his big bang theory, these three English astronomers published an opposing theory, which had this to say about time: The universe has no beginning and will have no end; hence, time always existed and will always exist. Time does not have a beginning.

How did this theory explain the receding galaxies? Simply by assuming that they were always there and always speeding apart from one another. They would continue to speed apart forever, for the universe went on indefinitely without boundaries. Moreover, as a galaxy sped away from one region of a space, another galaxy formed to take its place. As a consequence, the universe always looked the same; it was in a steady state. The theory advanced by Hoyle and his associates was called the steady state theory.

Scientists at first couldn't choose between the big bang theory and the steady state theory. Both appeared to be flawed. The opponents of the steady state theory pointed out that there was no evidence that particles, no less galaxies, could form out of empty space. Scientists who rejected the big bang theory did so on the grounds that there was no tangible evidence that a stupendous explosion had ever roared through the universe.

For scientists to start their journey through time from

time's beginning, the big bang theory had to be right. But where would they find evidence of an explosion that might have occurred billons of years ago and destroyed every bit of the primordial neutron glob? Gamov knew where.

The Echo of the Big Bang

According to Gamov, the universe resembled a white-hot fireball in the first moments after the big bang. As the universe expanded and cooled, the fireball gradually lost its brilliance. But, Gamov calculated, the radiation from that fireball should still remain. It is no longer visible because, moving away from us at enormous speeds, it has been red-shifted beyond the infrared into radio waves of short wavelengths known as microwaves. He considered this microwave radiation as the echo of the big bang.

In 1948, when Gamov predicted the echo of the big bang would be found, no instrument existed that could detect microwave radiation from space. But in the years that followed, radio astronomy developed rapidly. In 1964 a group of Princeton University scientists headed by Robert Dicke and P. J. E. Peebles predicted that the echo of the big bang could be detected with a radio telescope designed to pick up microwaves, and they began to build one. In the same year, a team of Soviet scientists made the same prediction, adding that an instrument with the capacity to observe microwave radiation from the big bang was already in operation at the Bell Laboratories in Holmdel, New Jersey.

The Holmdel instrument looked like a horn the size of a freight car, and scientists who worked with it called it the Bell Horn. The Bell Horn was not used for astronomical purposes, but rather to receive microwave messages from Bell's Telestar communication satellites. It was the job of two American electronic engineers Arno Perzias and Robert Wilson to bring in these messages as clearly as possible. That meant eliminating all sources of static in the Bell Horn itself. Penzias and Wilson did it, yet a low steady hiss persisted in the receiver. Clearly, this noise was coming from outside the Bell Horn.

The hissing sound came from all parts of the sky, day and night. The entire universe seemed to be its source. What was this mysterious noise from outer space? Penzias and Wilson were baffled. They were unaware that Soviet scientists had predicted the Bell Horn could pick up the echo of the big bang. Moreover, they had no knowledge of the work of the Dicke-Peebles team at Princeton, only 30 miles away from Holmdel. And neither the Russians nor the Princeton researchers knew about the results Penzias and Wilson had obtained.

By a lucky chance Penzias mentioned the mysterious hiss to a fellow scientist, who told him of the work the Dicke-Peebles team was doing. Penzias and Wilson drove the 30 miles from Holmdel to Princeton and swapped information. The Princeton team had calculated the kinds of wavelengths they expected to find in the echo of the big bang, and the pattern that Penzias and Wilson had found corresponded exactly. When both teams of scientists published their findings, Walter Sullivan wrote in a first page story in the *New York Times,* dated May 21, 1965:

"Scientists at the Bell Telephone Laboratories have observed what a group at Princeton University believes may be remnants of an explosion that gave birth to the universe."

No other way has since been found to explain the microwave radiation found by Penzias and Wilson. Since their discovery, "supporters of the steady state theory have tried desperately to find an alternate explanation," Robert Jastrow director of NASA's Goddard Institute of Space Studies, wrote in 1978, "but they have failed. At the present time, the big bang theory has no competitors."

And according to that theory, time did have a beginning. Science's journey from the beginning to the end of time could begin.

Science's time machine was a mental one, and it was made up of three components: The first was the scientist's knowledge of how matter and energy behave; and the belief that matter and energy have always behaved the same way and will always behave the same way. The second was the scientist's ability actually to see the past, for when we observe a galaxy at a vast distance, we are seeing it at the time light left it, millions to billions of years ago. And the third, and by far the most important component, was the scientist's ability to gather together all he knew and all he saw and, from the mass of apparently unrelated facts, to reconstruct what had happened in the past and to project what will happen in the future.

With that time machine, scientists—mainly astrophysicists, astronomers who are also physicists—made an enormous leap into the past to the moment time began. Then they started the long journey forward into time.

The Journey Through the Past

At the moment of the big bang neutrons and photons—elementary particles of matter and energy, respectively—were hurled into space at temperatures of trillions of degrees. In the first one millionth of a second of the universe, photons collided to form pairs of electrons and antielectrons, and antielectrons collided with neutrons to create protons and antineutrinos.

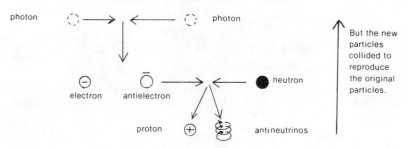

In the first one millionth second of creation, collisions of elementary particles produced new particles.

But the new particles collided to reproduce the original particles.

But matter and energy were packed so tightly together that almost instantaneously, the newly produced particles collided with one another to reverse the process of creation. Proton-antineutrino collisions produced neutrons and antielectrons;

antielectrons smashed into electrons and vanished in a puff of photons. The composition of the universe was much as it had been before the big bang.

But the universe was expanding and that changed everything. At the end of 1 second, there was so much space between particles that the chance of collision between photons decreased, and less electron-antielectron pairs were produced. There were now less antielectrons to combine with neutrons, and the number of neutrons began to rise. There was also less chance of proton-antineutrino collisions, and the number of protons began to grow as well. The stage was now set for the union of protons and neutrons. But temperatures were still exceedingly high, and that meant protons and neutrons sped through space at enormous energies. When they collided at these energies, they bounced off one another.

At the end of 3 minutes, the universe had expanded vastly. It was a gas; and when a gas expands, it cools. An immense temperature drop sharply cut the energies of the protons and neutrons. Now when these particles collided, they no longer rebounded; they stuck together to form an isotope of hydrogen called deuterium, or heavy hydrogen.

neutron   proton   deuterium
(heavy hydrogen)

The racing deuterium particles were now the heaviest in the universe, and protons and neutrons colliding with them were easily brushed aside.

But at the end of 3¾ minutes, further expansion had reduced temperatures to such an extent that the deuterium particles were no longer energetic enough to deflect protons and neutrons. They adhered to deuterium, forming the nucleus of helium.

A helium nucleus cannot pick up another proton or neutron without becoming unstable, so the building of more complex elements could not occur. The universe was now composed of 75 percent hydrogen and 25 percent helium.

After the first 3¾ minutes, the universe continued to expand and cool, but nothing important happened until 700,000 years had passed. At that time the universe had become so cool that electrons lost much of their heat-generated energy and were captured by protons to form hydrogen atoms, the first atoms to be created. Shortly afterward, electrons fell into orbit around helium nuclei, and helium atoms came into being. The buildup of hydrogen and helium atoms continued slowly.

When the universe was 200 million years old, the process began that would lead to the formation of quasars, galaxies, stars, and planets. The universe then consisted of huge thin clouds of hydrogen and helium atoms. By chance, more atoms of these clouds collected in some regions than in others. The force of gravitation then came into play.

This force attracts every particle in the universe to every other particle, and it is strongest when the particles are close together. It is however, the weakest force in nature, and the gravitational attraction between two neighboring atoms is so small as to be negligible. But the gravitational attraction of a great many atoms close together is strong. When large quantities of hydrogen and helium atoms collected in certain regions of a huge thin cloud, the gravitational force pulled the atoms together into small compact clouds.

The greater the amount of matter in a body—that is to say, the greater its mass—the greater its gravitational force. The massive compact clouds easily attracted more hydrogen and helium atoms as well as other compact clouds. The bigger, more massive clouds that resulted attracted even more matter and more compact clouds; and over the course of hundreds of millions of years, the compact clouds of hydrogen and helium atoms grew larger and larger. Quasars and galaxies formed out of these clouds; quasars between 1.5 to 3 billion years after the big bang, galaxies between 5.5 to 7 billion years.

Quasars were then at the edge of the expanding universe and have continued to be there ever since. Astrophysicists, journeying through the past, were unable to make out how quasars came into being, but they were able to obtain the following clear picture of the creation of galaxies.

The huge massive clouds of hydrogen and helium gas were in rapid motion, and that set up eddies—small concentrated whirlpools—around the edges of the clouds. As these eddies moved about, they collided with one another, and the gravitational force bound them together. The eddies grew steadily as a result of repeated collisions. When they reached a certain mass, the gravitational attraction between their own atoms became so great that most of the atoms were pulled together into the cores of the eddies, and the eddies contracted.

When a gas contracts, temperatures rise; and the temperatures of the eddies of hydrogen and helium gas soared. When the temperatures reached a certain point—the critical point, about 15 million to 21 million degrees Centigrade—a nuclear reaction started in the highly compressed cores. Four hydrogen nuclei (protons) underwent a series of transformations to produce a helium atom.

The mass of a helium atom is somewhat less than that of four protons. In the nuclear process by which the helium atom was formed—it's called a fusion process because the particles unite, or fuse, to form a new particle—some of the mass of the protons was converted into energy in accordance with Einstein's famous equation $E=mc^2$ (read E equals $m\ c$ squared). The equation means that when matter is converted into energy, the amount of energy released *(E)* is calculated by multiplying the amount of matter converted, the mass *(m)*, by the speed of light *(c)* multiplied by itself *(c^2)*. Since that latter number is enormous (186,000 times 186,000 equals 34,596,000), even a tiny amount of mass can be transformed into a great deal of energy.

When about .1 ounce of hydrogen is converted into helium, enough heat energy is produced to melt 20 million tons of ice.

When the atoms of millions of tons of hydrogen were converted into helium atoms in the cores of the compact eddies, a quantity of heat energy was produced so stupendous that the eddies began to glow with a blue light. In this way stars were born.

Stars formed around the edge of the huge gas clouds. Heavier than the clouds, they plunged inward toward the center, gathering more hydrogen fuel for the nuclear fires that continued to burn in their cores. Over the course of a few million years, billions of stars were created in a single gas cloud. When the process ended, all the gas had been used up, and the cloud had been transformed into a galaxy. Our galaxy was created in this fashion 8 billion years after time began.

The stars that formed the first galaxies are the oldest stars in the universe. They are called Population II stars. (The discovery of two different populations of stars, one older than the other, was made by the American astronomer Walter Baade in the early 1940s.) As the Population II stars burned up their hydrogen, they became hotter, bluer, and more luminous. Finally, they exhausted the supply of hydrogen in their cores, and the fusion process stopped.

The cores now consisted of a mass of helium atoms. The energy that had been produced by the fusion process had offset the gravitational force pulling the helium atoms together. With that energy gone, the helium atoms fell toward each other, the core contracted, and the temperature mounted. When the temperature reached a critical point—about 140 million degrees Centigrade—the nuclei of helium fused to produce an isotope of beryllium,

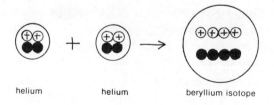

which combined with more helium nuclei to form carbon.

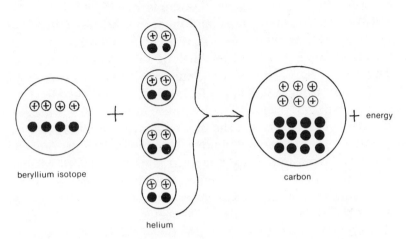

Since the mass of six protons was lost in the formation of a carbon atom, an enormous quantity of energy was generated. This started a nuclear chain reaction in the core of the stars, producing some of the heavier elements, including oxygen, nitrogen, and phosphorus, which with hydrogen and carbon would later form the basis of life. Temperatures at the cores of these stars were insufficient to trigger the heavy-element building process beyond iron.

The enormous amount of energy generated by the nuclear chain reaction shoved matter away from the cores of these Population II stars, and the stars expanded gigantically. The expansion cooled the surfaces, and the stars changed color from blue, which is produced by high temperatures, to red, which is produced by somewhat lower temperatures. These stars are known as red giants. The diameter of Betelgeuse, a typical red giant, is 300 million miles. It is 350 times as wide as the sun, which has a diameter of 865,000 miles. If Betelgeuse were to be put in place of the sun, it would occupy space beyond the orbit of Mars.

After million of years, the helium in the cores of the red giants became exhausted, and no energy was generated. The matter that the nuclear energy had pushed away from the core collapsed with sudden violence, causing the red giants to explode and become enormously brighter than they had been before. Those that became thousands of times brighter are called novae (singular nova), and those that became millions of times brighter are called supernovae. (The explosion of a star seen from earth looks like the birth of a new star. Hence *nova,* which is Greek for "new.") The energy released by a supernova explosion is 1,000,000,000,000,000,000,000,000,000 times greater than that released by a hydrogen bomb. The less massive red giants became novae; the more massive, supernovae.

After the explosion of a nova, the amount of mass that had not been blown into space remained large. The tug of gravitational attraction pulled it together, and the star shrank from a giant hundreds of millions of miles across to a dwarf the size of the earth, about 8,000 miles in diameter. The enormous mass now occupied such a small volume that a chunk of the star the size of a lump of sugar had a mass of a ton.

The atoms were crushed together so tightly in that small volume that they disintegrated. Elements ceased to exist. The star was composed solely of electrons, protons, and neutrons. The energy of the speeding electrons—multibillions of them—balanced the pull of the gravitational force and prevented the star from shrinking further.

Such a shrunken star had a high temperature as a result of the explosion and the subsequent contraction; it glowed with a white light, and it is called a white dwarf. But the star was not hot enough to set off a fusion reaction. Without a nuclear fire in its core, it would eventually cool, and the light produced by its heat would fade away. A white dwarf was a dying star. When it became a black dwarf, it would be dead.

After the explosion of a supernova, on the other hand, the shrinkage did not stop at the white dwarf stage. The mass remaining after an explosion of a supernova was far greater than the mass remaining after the explosion of a nova, and consequently so was its gravitational force. At the white dwarf stage that force overcame the resistance of the electrons, and the star continued to shrink until it was only a few miles wide. So small was the volume of this supershrunken star, and so stupendous was the mass in it, that a chunk of the star the size of a lump of sugar had a mass of 100 million tons.

The elementary particles in this star were packed millions of times closer together than in a white dwarf. The electrons were mashed into the protons and formed neutrons. The star became a mass of neutrons jammed surface to surface—a neutron star.

The neutron star did not collapse any further, possibly because the contraction had given it an enormously rapid spin, which generated sufficient energy to fight off the pull of gravity. The energy generated by the spin acted on the matter around the star as well—the debris of the supernova explosion—to produce radio waves and other radiation. The spinning star beamed out this radiation in pulses, much as a lighthouse sends out pulses of light rays. For that reason, neutron stars are also called pulsars. (It was from its pulsating radio waves that the first neutron star was detected by the English radio astronomers Jocelyn Bell and Antony Hewish in 1967. The following year a team of American astronomers found a pulsar spinning at a rate of 33 times a second in the Crab Nebula, where Chinese astronomers in the eleventh century had observed a supernova explosion.)

The neutron star glowed with a yellow light; but like the white dwarf, it too would cool down and dim into darkness. Its spin would slow gradually until it stopped, and when it did, it would cease to send out pulses of radiation. Like the white dwarf, the neutron star was a dying star. It, too, would become a dark, still, cold rock drifting in space.

As the shrunken ruins of Population II stars were dying—about 13 billion years after time began—new stars were being born out of the debris of the shattered red giants. What happened was this:

During the explosion of novae and supernovae temperatures soared colossally, and the nuclear buildup of heavier elements, which had stopped at iron, now continued. Bismuth, gold, lead, uranium, and other elements of high atomic weight were forged in those blasts. Then the mix of elements in the exploding red giants were flung out into space, enriching the galactic clouds of hydrogen and helium. Those clouds now contained all the elements that make up the universe as we know it, and from these clouds new stars were formed. These stars are called Population I stars, and they are still found in the outer regions of galaxies, where they were created. Our sun is a Population I star.

It was born 13.4 billion years after the beginning of time. Judging from its mass at birth, the scientific time travelers can foretell its future. It will swell into a red giant, swallowing the earth in its fiery interior as it expands, then explode into a nova and shrink into a white dwarf and die. But all this will not happen for many hundreds of millions of years.

At the same time the sun was born, its planets formed out of the same element-rich cloud.

The eddies in the galactic cloud condensed to form the sun and planets.

None of the planets had enough mass to generate sufficient gravitational contraction to start a fusion process. Jupiter had almost enough mass, and it just missed being a companion sun to our own.

Whenever a star was born—a Population I star or a Population II star—there was always a chance that planets would form with it. Scientific time travelers estimated that a hundred billion planetary systems were formed in our galaxy alone. But only the planetary systems of Population I stars could bring forth life. That was because the clouds from which Population I stars and their planets were formed contained the elements of life—carbon, oxygen, nitrogen, phosphorus, and hydrogen. The clouds from which Population II stars and their planets were formed contained only one of these vital elements, hydrogen.

This is how life was formed on earth (according to a theory that has its foundations in the work of the Russian biochemist A. I. Oparin published in 1924): Some of the vital elements—

carbon, nitrogen, and hydrogen—combined to form abundant amounts of the gases ammonia and methane in the atmosphere. There also was oxygen in the atmosphere, and about a million years after the birth of the earth, lightning flashes provided the energy to fuse ammonia, methane, and oxygen into amino acids, the building blocks of life. The amino acids fell to the ground in rain and collected in pools.

These pools contained phosphorus compounds. Colliding amino acids and phosphorus compounds stuck together and gradually, over millions of years, built an extremely long molecule in the form of a double helix (one corkscrew structure winding around another). That molecule, like no other molecule in nature, can make copies of itself. It is called DNA (deoxyribonucleic acid), and it is the stuff of life. Wrapped in a container of amino acids linked together (protein), DNA became a virus—the first living thing.

Within 8 million years from the time the first amino acids formed in the earth's atmosphere, new simple life forms, such as bacteria and algae, had developed from the original DNA. Then over more millions of years, a variety of plants and animals evolved. Manlike creatures appeared between 3 and 3.5 million years ago; and *Homo sapiens,* the race of man that populates today's earth, arrived on the scene about 250,000 years ago—18 billion years after the big bang that started time. (Allan Sandage, who had photographed the spectrum of the first observed quasar, calculated this age of the universe based on the rate of expansion of the galaxies. From the level of cosmic background radiation—the echo of the big bang—other scientists have deduced that the universe is 20 billion years old.)

From the birth of man to today, the universe has remained much the same. Stars have been born. Stars have erupted and died. The galaxies have sped farther and farther into space, and so have the quasars. At their stations on the rim of the expanding universe quasars are now at least 1 billion light-years beyond the edge of the observable universe. (The calculation is based on the rate of expansion of the universe.)

Astrophysicists had journeyed from the beginning of time to the present. Now they were ready to launch a mission to the end of time. But before they could start on their voyage, they had to answer a question much like the one they asked before they set off on their journey through the past:

Is There an End to Time?

Some astrophysicists answered no. They based their reply on the assumption that the universe would never stop expanding. Space would become emptier and emptier as the galaxies flew apart. Stars would die out one by one. The material for new stars would become exhausted. Black galaxies would speed out into a black universe forever. The universe would be dead. But time—since it exists when things change, even when they only change their positions—would never end.

Other astrophysicists, led by Allan Sandage, answered yes. They based their answer on the fact of escape velocity. If a space rocket is to leave the earth, it must have a certain minimum velocity, an escape velocity. If this velocity is not reached, the rocket will slow down and fall back to earth, pulled there by the gravitational force of the earth's mass.

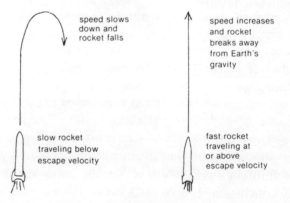

Similarly, if galaxies are to continue to speed apart, they must have an escape velocity, a speed great enough to overcome the gravitational force of the universe's mass. From a study of red shifts, Sandage discovered that some galaxies were slowing down—a strong indication that they did not have an escape velocity. From this, Sandage drew this picture of the future of the universe:

The galaxies will fall back toward each other with greater and greater speed, finally meeting in one region of space. The total mass of the universe close together will generate a gravitational force so great that all the galaxies in the universe will be pulled together into a globe smaller than a neutron star. Jammed into that tiny space, all the matter in the universe will be crushed into elementary particles—the jam of neutrons and photons with which the universe began. The universe will end as a primordial atom in which nothing changes. Time will come to an end.

If Sandage was right, then there has to be enough mass in the universe to generate sufficient gravitational force to slow down the galaxies. Is there? In 1974 a team of sixty-five American astrophysicists headed by J. Richard Gott III estimated the mass of all the matter in the galaxies and between the galaxies—the total mass of the universe—and answered: There is not. As a matter of fact, Gott reported, the actual mass of the universe is only 10 percent of that necessary to slow down the galaxies.

Sandage's followers asked: Where is the rest of the mass of the universe? And some of those astrophysicists answered: Hidden where no man could ever see it—in the most bizarre objects in the universe:

The Black Holes in Space

The story of the black holes begins in 1916 when the German physicist Karl Schwarzschild solved some of the equations of Einstein's general theory of relativity and came to this startling conclusion: If a star of a certain mass is compressed to a certain radius—the Schwarzschild radius—the star would collapse to a volume of less than a few cubic miles.

The concentration of a star's enormous mass in that small volume would produce a gravitational force so great that the escape velocity would exceed that of light. That meant that nothing could escape from this supershrunken star—not even light. The star would be forever black. And since whatever fell into it, matter or energy, could never emerge from it, the star would act like a bottomless hole—a black hole in space.

In 1938 the American nuclear physicist J. Robert Oppenheimer demonstrated mathematically that the most massive of the red giants—red supergiants—could collapse through the white dwarf and neutron star stages to objects with radii smaller than the Schwarzschild radius. These objects would vanish from sight and become black holes.

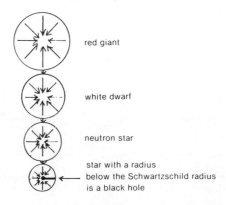

Could black holes contain the necessary 90 percent of the mass of the universe necessary to account for the slowing down of galaxies Sandage had reported? Possibly. But since black holes were invisible and prevented radiation of any kind escaping from them, how could they be observed? And if they couldn't be observed, who was to say that they were not, as one scientist put it, "of all the conceptions of the human mind from unicorns to gargoyles . . . the most fantastic?"

But though black holes couldn't be observed directly, they could be detected by their action. If another star revolved around a black hole, then the intense gravitational pull from the black hole would draw gas from the surface of the star. The gas would swirl around the black hole in the form of a disk before being sucked in. This gas would be compressed and speeded up tremendously by the gravitational force of the black hole and would emit X rays.

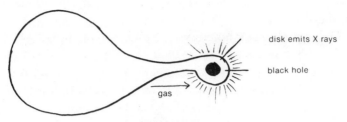

The shape of the star is distorted by the gravitational pull of the black hole.

If a star could be found orbiting around an X-ray source, the source would be a black hole.

X rays from space cannot be detected on earth because the atmosphere shuts them off. But they can be detected by X-ray telescopes in satellites flying above the atmosphere. In 1971 the satellite *Uhuru* spotted the X-ray source Cygnus X-1 (the name means the first X-ray source found in the constellation Cygnus). Optical telescopes on earth identified a supergiant star orbiting Cygnus X-1. Cygnus X-1 was, therefore, a black hole—the first to be discovered.

Cygnus X-1 is several times more massive than the sun. (The sun has a mass of 2,000 billion billion tons—2 followed by 21 zeros.) More massive black holes were thought to be located in Seyfert galaxies, a class of galaxies within which violent and explosive events of unknown nature have occurred from time to time.

Drawing on observations of *Uhuru* and of the British X-ray satellite *Ariel 5,* a team of British astronomers determined that some Seyfert galaxies are ten times brighter in X rays than in visible light. The X rays come from the core of the galaxies. The American astronomer Harvey Tannenbaum speculated that there are black holes at the cores ten to a hundred million times more massive than the sun, sucking in not just gas from stars but whole stars.

In April 1978 the American astronomer Jerome Kristian published X ray evidence of a black hole five billion times more massive than the sun in the core of a Seyfert galaxy known as M87. Our own galaxy, according to findings announced in March 1979 by the American astronomer Eric J. Chaison, contains a black hole at its core five million times more massive than the sun.

Do enough black holes exist to make up the missing 90 percent of mass necessary to explain Sandage's observation of slowing galaxies? Not likely, according to Gott's research team. Moreover, some astrophysicists believe that black holes do not exist at all, despite the X-ray evidence. They point out that according to today's interpretations of Einstein's equations, the cores of black holes are supposed to continue to shrink to no volume—nothing—and yet that nothing is supposed to contain an infinite amount of matter—a greater amount of matter than can be thought of. Since that description has no relation to any observed object, these astrophysicists believe black holes are no more than a mathematical fiction, and that the X-ray sources now identified as black holes will turn out to be something quite different.

If black holes do not supply the complete missing mass, Sandage's followers argued, they supply some of it; and certainly more mass will be found somewhere in the universe. In its first four months of operation, the American Einstein satellite, launched in January 1978, found evidence of a vast quantity of gas glowing with X rays in the region of the most distant quasars. The gas may be sufficiently extensive to provide the rest of the mass needed to slow down, then ultimately reverse, the flight of the galaxies.

If it is, then Sandage, with his picture of the future of the universe, has already journeyed to the end of time. The date of his journey's end? According to Sandage, 67 billion years from now—85 billion years from the moment time began.

But the journey through time is far from over. Vast areas of the unknown remain to be explored.

How were quasars born? No nuclear reaction can generate their tremendous energy; is there a powerful unknown energy source in the far reaches of the universe? What happens to matter and energy when they are swallowed by a black hole? Are they compacted into something like a primordial atom? If so, does the black hole explode and create a new universe? Did our universe begin as a black hole? Instead of exploding, does the matter and energy in a black hole worm through space and emerge elsewhere in our universe (or in another universe) as a white hole—a source of unimaginable violence?

The American physicist Richard Feynman holds that antiparticles can be viewed as normal particles moving in reverse time; does antitime as well as antimatter exist? Is there, as Richard Gott, who conducted the survey of the amount of matter in the universe, suggests, an antitime region in our universe where time flows backward? Does time, as Einstein's equations predict, come to a stop in a black hole?

When in the distant future the universe contracts into a primordial atom, will it explode again? Then expand, contract, explode again; and will that cycle continue endlessly? If there are successive universes, will each evolve in the same way, so that history will always repeat itself? Or will wildly different kinds of universes be born each time with, as the American physicist John Wheeler suspects, alien geometries and undreamed-of physical laws?

But of all the unsolved mysteries the most profound is: What brought the universe into being? In the beginning according to *Genesis,* there was nothing but darkness and the void. Then God said, "Let there be light," and light appeared. In science's version of genesis, in the beginning there was also nothing but darkness and the void—the primordial atom and space—until the big bang set the universe ablaze with radiation. In *Genesis,* God was surrounded by chaos until in six days he created the orderly structure of the universe. In the first $3\frac{3}{4}$ minutes in science's genesis, the fundamental particles of energy and matter and their orderly relationship to each other came into existence.

Are these parallels significant? Albert Einstein, viewing the universe, was "astonished to notice how sublime order emerges from... chaos." On another occasion he wrote, "I believe in [a] God who reveals himself in the orderly harmony of what exists." Perhaps, at this point in science's long journey to find the secrets of the universe through reasoned inquiry, the road turns to faith.

Index

Alpha Centauri, 120-24
Alpha rays
  composition of, 69, 82
  discovery of, 64-67
  as nuclear exploratory tool, 83-84
Anderson, Carl, 93
Antimatter, 93-96
Atlantic Ocean
  history of the, 20
  map of the floor of the, 13
Atom(s)
  ancient Greek concept of, 57-58
  birth of, 172
  discovery of, 61 passim
  map of, 80, 87
  size of, 63, 83
Atom smashers, 90-91
Atomic weights, 68

Baade, Walter, 174
Becquerel, Antoine Henry, 63-64
Bell Telephone Laboratory
  (Holmdel, N.J.), 150, 168
Bessel, Friedrich, 125
Beta rays
  composition of, 77
  discovery of, 64-67
Big bang, 166, 168
  echo of the, 168-69
  first 3¾ minutes of the, 170-72
Bjorken, James, 106-8
Black holes, 163, 181-82
Blinking stars. *See* Cepheids
Bohr, Niels, 81-82
Bondi, Herman, 167
Bothe, Hermann, 83
Bottled lightning experiments, 73-76
Brightness, stellar

  absolute, 126
  apparent, 127
  relationship with distance, 130
Bubble chamber, 100

Calm coast, the mystery of the, 44-45
Cepheids, 126
  period-luminosity relationship, 126, 128 passim
CERN, 99
Chadwick, James, 84
Chaison, Eric J., 184
Clarke, Arthur C., 118
Cloud chamber, 91-92
Coalsack Bluff, 50
Colbert, Edwin H., 50
Color force, 104-7
Comets, 10-11
Compass needles
  clue of the misdirected, 46-47
  flip-flopping, 16-21
Continental drift, 34, 49
Convection currents, 19-21
Cosmic rays, 96
Cowan, Clyde, 98
Curie, Marie, 64-65
Curie, Pierre, 64

Dalton, John, 59-63, 67-68, 70
Deep-sea zebra stripes, puzzle of the, 35-38
de Sitter, Willem, 163
Deuterium (hydrogen isotope), 99, 171
Dicke, Robert, 168-69
Dietz, Robert S., 48
Dirac, Paul, 93-94, 96

Distances, galactic
  by the red shift method, 144 passim
  See also Cepheid and red giant methods *under* Distances, stellar
Distances, stellar
  by the Cepheid method, 128 passim
  by Henderson's method, 119-24
  by the red giant method, 136
DNA, 179
Doppler effect, 142-44
Doppler, Johann Christian, 144
Doyle, Arthur Conan, 28

Earth
  core, inner, 25
  core, outer, 24
  cross section of the, 21
  crust, inner, 7-8
  crust, outer, 7-8
  mantle, 45
Earthquakes
  causes of, 42-43
  in Kulpa Valley, 5-8
  in ring of fire, 41-43
  squiggles of, 3-4
  waves of, 3, 22-23
Echo sounding, 12, 30-31, 42
Eddington, Arthur Stanley, 164
Einstein, Albert, 163 passim, 185
  his equation relating mass and energy, 173
Electric force
  in atom, 88-89
  defined, 73
  nature of the, 77
Electromagnetic radiation, 148-50
Electron
  in atomic shell, 77-78
  at birth of universe, 170
  discovery of the, 73-77
  motions of the, 81
Elements, 58-59
Escape velocity, 180
Ewing, Maurice, 39
Expanding universe, 163 passim

Fermilab, 104
Feynman, Richard, 185
Friedman, Alexander, 164

Galaxy(ies)
  definition of, 138
  discovery of our, 136-38
  expanding, 141 passim
  formation of, 173-74
  other than ours, 139 passim
  shape of our, 138
  size of our, 138
Gamma rays
  composition of, 77
  converted to matter, 95
  discovery of, 64-67
Gamov, George, 166-68
Gell-Mann, Murray, 101-3, 107
Genesis, 185
Ghost particle, 96-98
Glashow, Sheldon, 107-8
Globular clusters (globulars), 136-38
Glomar Challenger, 46-47
Glossopteris, 29-31
Gluons, 104-6
Gold, Thomas, 167
Gondwanal, 30, 41
  map of, 48
Gott, J. Richard, 181, 185
Gravitational force, 172 passim

Hadrons, 100 passim
Halley, Thomas, 12
Hasse, Henry, 57
He Who Shrank, plot of, 55-57
Heezen, Bruce, 39
Helium nucleus, 83, 171-72
  as source of stellar energy, 174-75
Henderson, Thomas, 120-24, 131
Hertz, Heinrich Rudolph, 149
Hertzsprung, Ejnar, 126-34, 136
Holden, John C., 48
Hot ice, 15-16
Hoyle, Fred, 167
Hubble, Edwin P., 139-40, 144-46, 165
Hubble's law, 145
Humason, Milton, 144, 152

Index

Isotopes, 86

Jansky, Karl, 148, 150-51
Journey to the Center of the Earth, The, plot of, 1-3

Kant, Immanuel, 139
Kristian, Jerome, 183
Kubrick, Stanley, 118

Lamont-Doherty Geological Observatory, 12, 39
Land bridges, mystery of the missing, 29-31
Leavitt, Henrietta Swan, 124-26
Lederman, Leon M., 109-10
Lemaître, Abbé George, 165
Leptons, 99
Life, birth of, 177-78
Light buckets, 147. *See also* Telescope
Light waves, 141
Light year, definition of, 124

Magnetism
 definition of, 88
 ocean floor, 35-40
Magnetometer, 35
Magnitude, stellar, 127-28. *See also* Brightness
Mallet, Robert, 3
Maracot Deep, The, plot of, 27-28
Marconi, Guglielmo, 149
Matthews, Drummond, 35-38
Maxwell, James Clark, 148-49
McKenzie, Dan, 42, 44
Meson(s)
 discovery of, 89, 93
 model of a, 103
Messier, Charles, 140
Mid-Atlantic Rift, 13, 16, 32-33, 34-35, 39
Minkowski, Rudolph, 152
Mirror-image pattern, clue of the, 39-41
Moho, 11
Mohorovicic, Andrija, 3-9
Morley, L. W., 16-17, 39
Mount Palomar Observatory, 147

Mount Wilson Observatory, 136, 139
Muon, 96

Nebula, 139. *See also* Galaxy
Neutrino, 98
 at birth of universe, 170
Neutron(s)
 at birth of universe, 170
 discovery of, 83-86
Neutron star, 176-77
Nova, 176
Nuclear energy, 73
 in stars, 173-75
Nucleus, atomic
 discovery of, 70-73
 size of, 83

Ocean floor
 magnetism, 35-41
 spreading, 19-20, 38-39
 tracks of the moving, 46-47
Oparin, A. I., 178
Oppenheimer, J. Robert, 182

Pangaea, 48
Parallax
 definition of, 121
 method of determining stellar distances, 122-25
Particle accelerators. *See* Atom smashers
Pauli, Wolfgang, 96-98
Peebles, P. J. E., 168-69
Penzias, Arno, 168-69
Perfect proportions, puzzle of the, 58-63
Photons, 88
 at birth of universe, 170
Plate tectonics, 49
Plates, global, 41
 direction of, 45-47
 future directions, 49
Positron, 93
Potassium-40 clock, 31
Primordial atom, 166, 185
Project Mohole, 10-12

Proton(s)
  at birth of universe, 170
  discovery of, 78-79
Psi/J particle, 106
Pulsars, 177

Quarks, 101 passim
  bottom, 109
  charmed, 106
  down, 101
  strange, 102
  top, 110
  up, 101
Quasars, 155
  formation of, 173

Radio
  "stars", 151 passim
  telescope, 151
  waves, 149-50
  waves from neutron stars, 177
Radioactivity, 64
  discovery of, 63-64
Radium
  decay into radon and helium, 69
  discovery of, 64-66
  gun, 66
  use of in discovery of alpha, beta, and gamma rays, 66-67
  use of in discovery of atomic nucleus, 70-72
Red giants, 175. *See also* Stellar distances
Red shift, 141 passim
Reiner, Fred, 98
Relativity, 163 passim
  and black holes, 181, 185
Rift-ridge system, global, 40
Ring of fire, 41-43 passim
Roentgen, William Conrad, 63
Rutherford, Ernest, 66-67, 69-73, 77-78

Sandage, Allan, 152-53, 170-80, 182, 184
Schmidt, Maarten, 152 passim
Schwarzschild, Karl, 181
Schwarzschild radius, 181-82

Scripps Institution of Oceanography, 39, 46
Seismograph, 4, 23-24
Seyfert galaxies, 183
Shadow zones, 22-26
Shapley, Harlow, 136-40
Sky, as seen in two and three dimensions, 134-35
Slipher, Vesto Melvin, 144-45, 164-65
Small Magellanic Cloud, 124 passim, 134
Soddy, Frederick, 67, 69
Solar system, 118-20
  birth of, 178
Spark chamber, 100
SPEAR, 108
Spectrum
  electromagnetic, 148
  infrared, 148, 153-54
  standard, 142
  visible light, 141
Stars
  birth of, 173-74
  distinguished from planets, 188
  Population I, 177
  Population II, 174
Steady state theory, 167
Stone icicles (stalactites), 9
Strange particles, 102
String of light, clue of the, 87-89
Strong force, 89
Struve, Georg Wilhelm von, 125
Superglobular, 137
Supernova, 176

Tannenbaum, Harvey, 183
Telescope, optical, 121, 146-47
  radio. *See* Radio
Telescopic photography, 146, 147-48
Tharp, Marie, 12-13, 40
Theory, scientific, 163
Thomson, J. J., 73-77 passim
Time
  beginning of, 169
  end of, 180 passim
  explanation of, 167
Time Machine, The, plot of, 157-62
Ting, Samuel, 110

Tracks of ocean plates, 45-47
Trenches, deep-sea, 42
Tsunamis, 43
2001, A Space Odyssey, plot of, 113-18

Universe
  date of birth, 179
  date of death, 184
  Shapley's, 138
Uranium
  first separation of pure, 64-65
  in Becquerel's experiments, 63-64

Verne, Jules, 3
Vine, Fred, 35-38 passim
Volcanic islands, 14, 31-35 passim
Volcanoes
  along Mid-Atlantic Rift, 14
  causes of, 43
  undersea, 14

Weak force, 99
Wegener, Alfred, 33-34
Wells, H. G. (Herbert George), 162
Wheeler, John, 185
White dwarfs, 176
Wilson, C. T. R., 91
Wilson, J. Tuzo, 13-21 passim, 32, 37, 45
Wilson, Robert, 168-69

X rays
  discovery of, 63
  stellar sources of, 183

Yukawa, Hideki, 87-92 passim

Zweig, George, 101-3, 107